THE OIL CENTURY
From the Drake Well
to the Conservation Era

The
OIL CENTURY

From the Drake Well
to the Conservation Era

by J. Stanley Clark

NORMAN : UNIVERSITY OF OKLAHOMA PRESS

By J. Stanley Clark

Open Wider Please (Norman, 1955)
The Oil Century:
From the Drake Well to the Conservation Era (Norman, 1958)

LIBRARY OF CONGRESS CATALOG CARD NUMBER: 58 : 11609

Copyright 1958 by the University of Oklahoma Press, Publishing Division of the University. Composed and printed at Norman, Oklahoma, U.S.A., by the University of Oklahoma Press. First edition.

To my friend

A. A. ("Red") Lynn,
a drilling contractor

FOREWORD

T HE AMERICAN petroleum industry marks its beginning with the completion of the Drake well near Titusville, Pennsylvania, in 1859. Within a decade the pattern of its development was set: exploratory operations in new territory, advanced techniques in drilling and transportation, expanded markets, improved refined products, new domestic and industrial uses.

Through the century of development its growth has kept pace with industrial demands.

Governor John F. Simms of New Mexico, chairman of the Interstate Oil and Gas Compact Commission, expressed this thought in his annual address, in June, 1956:

> This industry has overcome countless technical obstacles from the time the first well was spudded. Each new pool has presented its share of problems to be solved—and an alert industry, step by step, has solved them.
>
> A well might "crater" near Corpus Christi, blow out at Monument, threaten to choke itself in the "heavy shale" formations on the Gulf Coast, or get 11 degrees off in the chert in West Texas—but somehow, somebody always devised an answer to the problem and the industry moved on.

Similar adaptability has marked advancements in transportation, in the refining and petrochemical industries, and in the continuous search for additional uses and markets for petroleum products.

Consider the pipe-line industry, which marked its beginning with iron tubing of small bore permitting gravity flow for short distances. Contrast the present maze of feeder and trunk lines of thin-walled steel composition, which transport more barrels of crude oil and products in a week's time than were produced in any year of the nineteenth century. Contrast, too, the expansion in the industry of automation and microwave systems which regulate the pumping operations.

Or consider the refining industry—what a challenge a drop of oil has been, whether it is paraffinic, naphthenic, or asphaltic in molecular composition. In the beginning, there was the basic process of distillation—the heating of crude oil to distill off products. One of these, gasoline, was considered a waste product until the internal combustion engine ushered in the automobile age. Experiments in time developed the cracking process, the subjection of heavy or high boiling materials of large molecules to heat and pressure to break out lighter products such as gasoline. From this thermal-cracking process, the industry advanced to catalytic-cracking at lower pressures and lower temperatures to produce higher octane gasoline and a higher liquid recovery. Subsequently, there followed the process of reforming to improve the octane number or quality. In addition, alkylation, hydrogenation, and platforming have been developed for producing high octane gasoline.

Marked improvements have been made in the drilling industry since World War II: better bits, improved hydraulic systems and mud programs, greater mobility of drilling rigs, and the development of slim-hole rigs. And, as the first century of oil draws to a close, three new methods of drilling—"sonic," "percussion," and "turbodrill," which employ self-contained units at the bottom of the hole thus reducing energy losses from surface to bottom—are beyond the experimental stage.

New depth records in drilling have been set. A well was drilled in southern Louisiana in 1955 to a depth of 22,570 feet.

Ninety-nine wells in the country were producing from depths below 15,000 feet, seventy of these in southern Louisiana.

Exploration is a costly gamble. Based upon 1953 drilling costs, it cost $20.54 per foot for footage drilled between 7,501–10,000 feet, $41.80 per foot between 10,001–12,500 feet, $57.68 per foot between 12,501–15,000 feet, and $105.91 per foot below 15,000 feet.

Not only is exploration highly speculative and costly, but production does not necessarily assure great profit or wealth. When Drake "tapped the mine," he was under contract to pay a royalty of ten cents a gallon on all crude oil produced; in too short a time Pennsylvania crude was selling for ten cents a barrel. Periods of "feast and famine" have been too common. At the turn of the century, Spindletop came in with a roar and the price of oil fell to three cents a barrel, less than the going rate there of a dipper of water. A few years later the San Joaquín field in California tumbled prices to ten cents a barrel. The gushing abundance of the Cushing field in Oklahoma in 1915 forced prices downward to twenty-three cents. Two years later World War I demands for Cushing oil skyrocketed prices to four dollars a barrel. The Oklahoma City field and the East Texas field in 1931 and again in 1933 plunged prices downward to ten cents a barrel. And price controls in the national emergencies of 1942–46 and 1951–53 held prices below a spiraling wholesale commodity index.

When the Sixteenth Amendment to the federal Constitution —the income tax amendment—was adopted in 1913, American industry and representatives in the national Congress, through public hearings, helped formulate a revised revenue code. Congressmen and members of the oil industry alike were concerned with new and untried principles of taxation.

In order to encourage exploration to meet increased demands for petroleum products in the rapid development of the motor age, and to take care of the depletion of a natural resource, the federal government provided for a "reasonable allowance" for this purpose in the Revenue Act of 1918. Officials of the Treasury Department charged with administering the act, adopted a percentage depletion of 27½ per cent. Section 114(b)(3) of the Internal Revenue Code of 1939 provided that percentage depletion should be a deduction equal to 27½ per cent of the gross income for the producing property, but this should not exceed 50 per cent of the net income from the property. The provisions have remained unchanged, although in recent years they have been under attack by some congressmen investigating integrated oil companies. By channeling the flow of funds into crude oil exploration and producing activities, a company can modify the impact of an excess profits tax to a considerable degree.

Oil is a decisive factor in international power politics. It played a determinative role in the final decision of World War I; it was a major factor in forcing the outcome of the last great world conflict. Substratum areas of the earth where proved oil reserves exist continue to be of international concern. It is significant that within the past year—another year of the cold war but no armed conflict—more than one billion dollars were spent by the Department of Defense for aviation gas and jet fuels to power planes and bombers in a containment policy toward Communism.

More than two thousand years ago the Greek historian Thucydides wrote: "We both alike know that into the discussion of international affairs the question of justice only enters where the pressure of necessity is equal, and that the powerful exact what they can, and the weak grant what they must." Americans within the past twenty years have become more responsive to the exigencies of world politics. Little more than a decade and a half have passed since Franco sent a congratulatory telegram to Tojo at the time of the fall of Corregidor and the death march of Bataan. Today, Spain and Japan are considered within the periphery of defense for our containment policies.

Every school boy in America knows how vital oil is to the security of the free world. The May 2, 1957, issue of *The Stars and Stripes* contains an article that points up the vital interdependence. The news story on the $400,000,000 U. S. military construction program in Spain described the 485-mile long Air Force multiproduct pipe line to transport jet fuel, diesel fuel, and motor and aviation gasoline. Underground tank farms along the route can store 6,000,000 barrels of fuel and the pipe-line capacity is 286,000 barrels. The need for a pipe line to sustain the jet bases in Spain was pointed up by the poor state of the Spanish highways and the fact that a wing of B–47 medium bombers consumes more fuel in one afternoon than the entire Spanish railway tanker fleet could transport in a month.

This is typical of the demands made upon the American petroleum industry by perimeter defense problems. During the first century of oil, the petroleum resources of no other country have been so heavily drawn upon to meet foreign needs as those of the United States. And, as the containment policy toward Communism has brought into sharper focus petroleum for defense purposes, average daily production in the United States has fallen behind daily consumption, and the need for a national policy toward the importation of oil and by-products has become increasingly apparent.

This does not imply that the United States has become a "have-not" nation, but state restrictions relative to physical waste and market demand curtail production. Interwoven with the ever present rivalry for market between stripper-well owners and other producers with limited production, and integrated companies with national and international holdings, is the problem of whether to produce irreplaceable national resources at a more rapid rate of exhaustion or to increase dependence on foreign supply.

Actually, a case might be made for the statement that interest in conservation practices quickened with the need for the importation of oil to meet the nation's demands in the middle 1920's. The principal similarity, nevertheless, was the time of occurrence. By the early thirties, there was growing acceptance of the idea in oil-producing areas that state legislation and enforceable orders by regulatory agencies were necessary to curb physical waste in the production of oil. In the petroleum industry, conservation practices in production are primarily concerned with the prevention of waste (physical waste below and above the ground). Although the Petroleum Experiment Station of the Bureau of Mines furnished the industry evidences of wasteful practices that could be corrected, the federal government was limited in its authority for direct control over public lands and leases of Indian wards.

The depression and flush production in the 1930's hastened co-operative efforts toward conservation, as petroleum engineers gained knowledge of reservoir energy and methods of production for increased recovery. This was a period of crisis in the oil industry: when oil-producing states experimented with diverse conservation regulations; when court battles were fought; when proration was enforced in some states while proration orders issued by regulatory commissions were ignored in many producing areas; when the very word "proration" became anathema in other states and is even now deliberately ignored in the conservation statutes of certain states; when a judge in a dissenting opinion from a State Supreme Court decision exclaimed: "In my opinion, proration of oil was born of monopoly, sired by arbitrary power, and its progeny (such as these orders) is the deformed child whose playmates are graft, theft, bribery, and corruption."

Out of that hectic period, oil-producing states evolved conservation legislation fully tested and sustained by state and federal courts. The region of greatest production, the Gulf Coast–Southwest and the Mid-Continent area, the seat of legislative battles on conservation legislation and the scene of set-

backs and triumphs from court decisions on administrative practices, furnished leadership that led an association of states to
form the Interstate Compact to Conserve Oil and Gas, in 1935.

The Compact represented compromise action. Governor E.
W. Marland of Oklahoma advocated the limitation of supply
to demand among the states, an intentional price-fixing device,
and sought participation of the federal government in order to
control and limit the importation of oil. Governor James V. Allred of Texas was strongly opposed to his state's entering any
compact which had price-fixing as a primary purpose. He advocated the prevention of physical waste. He did not propose any
encroachment by an interstate agency on the power of Texas
to control its petroleum production.

The Texas point of view prevailed. Texas has produced more
than one-third of the oil produced in the United States since
the Drake well and, since 1935, the year of the Compact, has
averaged more than 40 per cent of the country's daily production. Without Texas, any oil and gas compact between states
would have been ineffective. That state has more than one-half
of the nation's oil reserves. It has four-fifths of the nation's flowing wells, all under proratable production. Because of conservation practices that have protected reservoir energy, Texas
bore the brunt from, and was able to meet, increased demands
for production imposed by World War II, between 1941 and
1945, and the Suez Crisis, in 1956. A case might be made for
the point that the degree of stabilization reached by the petroleum industry is directly attributable to conservation practices
in Texas.

The Compact is an educational device which now includes,
with the exception of California, all the states in which oil and
gas are produced. It functions through the Interstate Oil and
Gas Commission, which includes a representative from each
member-state, and is solely supported by state contributions.

The Commission is a fact-finding body chiefly interested

in publicizing oil and gas conservation policies. This it accomplishes through semiannual meetings, attended by representatives of the industry and state and federal agencies interested in conservation, and by its publications. Its Legal Committee has drafted model conservation legislation useful as a guide for newer members of the Compact. Its Regulatory Practices Committee has prepared comprehensive drafts of rules and regulations and uniform reporting forms for the use of state regulatory bodies. An Engineering Committee has issued reports on production engineering. A Secondary Recovery and Pressure Maintenance Committee has issued studies in these related fields.

By Congressional direction, the Attorney General of the United States recently made an investigation of all activities of the Commission. The summary of findings, published September 1, 1957, includes these statements: "That a certain amount of general uniformity has developed in the States on conservation laws, rules and regulations is a tribute to its reputation and influence, not its power. . . . On the whole, the activity of the Commission appears to have been worthwhile. It seems justifiable to ascribe a good deal of the improvement in industry operations over the past quarter century—the elimination of gushing wells and flaring gas, the better use of reservoir energy, the rise in production of oil from about 20–40% to 80% of the potential of the well through utilization of advanced recovery practices— to the promotional activities of the Compact Commission. Above all, the Compact and its Commission are unique examples of effective cooperation on a wholly voluntary basis."

During the past quarter-century, the petroleum industry has overthrown an earlier practice of unlimited production and observance of the rule of capture, the drilling of too many wells that too quickly depletes reservoir energy. Substituted has been the belief and practice that any well which will not materially increase the total ultimate recovery from an oil or gas field is an unnecessary well. The closing years of the first century of oil witness increased activity in secondary recovery methods, the revival of apparently depleted fields, and deeper and more costly exploratory efforts to build up the nation's reserves.

J. Stanley Clark

OKLAHOMA CITY
AUGUST 25, 1958

ACKNOWLEDGMENTS

THIS STUDY was made possible through the co-operation of the University of Oklahoma Press, Norman, and the Interstate Oil Compact Commission, Oklahoma City. The complete files and records of the Compact Commission were made available for my examination by Judge Earl Foster, general counsel, and his assistants, Lawrence Alley and Albert E. Sweeney, Jr.

Ralph Hudson, the state librarian of Oklahoma, and his staff granted me unlimited access to the federal-state depository of records.

Material on the early beginnings of the oil industry was compiled during a five weeks' visit in Titusville, Pennsylvania, examining records housed in the Drake Museum. Grateful acknowledgment must be made to Paul H. Giddens, president of Hamline University, who had thoroughly sifted these records and thus made my search less difficult.

Particular thanks are due to Archie Brown of the Drake Museum and to Jim Stevenson, editor of the Titusville *Herald,* for the courtesies they extended to me, and to J. D. Sullivan, who resides at the site of Pithole City and who was kind enough to take me to many points of interest along Oil Creek. I should also like to thank Mr. Robert E. Hardwicke, of Fort Worth, for his aid in connection with illustrations.

The personnel of the American Petroleum Institute, Standard Oil of New Jersey, and the Cities Service Company were most helpful when I visited their offices in New York. Friendliness and co-operation, too, were my lot on visits to the Petroleum Experiment Station in Bartlesville, Oklahoma, for the examination of records of the Bureau of Mines.

Lastly, special appreciation is extended my typist, Miss Leona Mahaffey.

J. Stanley Clark

CONTENTS

ILLUSTRATIONS

THE OIL CENTURY
From the Drake Well
to the Conservation Era

1

THE CENTURY
BEGINS IN PENNSYLVANIA

S EVERAL remotely related incidents led to the beginning of the oil industry in Pennsylvania, which is to say, the United States. Visitors and explorers along Oil Creek in the 1700's took note of the seepages from springs along or in the bed of the creek. Indians periodically visited the area and dug pits in seepage spots to retain water and oil in order to skim the lighter oil from the surface. Settlers in the region obtained oil in small quantities by "merely digging holes in the ground, and when they were filled with oil and water they spread a piece of flannel or woolen cloth upon the oil, which rises on the water and is taken up by the cloth which is then removed and the cloth wrung out."[1]

The raw product had little local, primitive use: Indians used it to daub themselves; settlers used it for medicinal dosage, crude lubrication purposes, and occasionally illumination. The product, when burned, produced noxious odors and more smoke than light. In the first quarter of the nineteenth century, Indians from the vicinity of Oil Creek, a tributary of the Allegheny River, would occasionally visit Pittsburgh, their canoes filled with furs and other forest products, including oil stored in gourds, cala-

---

[1] From Colonel E. L. Drake's story, written in 1870. The manuscript is in the Drake Museum, Titusville, Pennsylvania. It appears in *Pennsylvania Petroleum, 1750–1852, a Documentary History,* compiled and edited by Paul H. Giddens (hereafter cited as *Pennsylvania Petroleum*). Giddens collected 172 documents which include copies from manuscripts, newspapers, and books published during the period.

bashes, or hollow logs fashioned into kegs. Merchants took it in exchange for goods and kept it on sale as a kind of universal panacea. It was considered especially useful as a salve for burns, scalds, and wounds or sores of both men and animals. It was called Seneca Oil.

Oil springs were known to settlers in New York, Ohio, western Virginia, and Kentucky, and to the Spanish settlers of California and Texas. In 1855, A. D. Searl, a surveyor, found oil oozing from the earth near the site of Paola, Kansas Territory, and in 1860 a shallow well found production. Oil springs were common in the lands granted to the Choctaw and Chickasaw tribes in Indian Territory, and in 1859 an unsuccessful attempt was made to produce oil in the Choctaw Nation by a company organized for that purpose.

In drilling salt wells, too, petroleum had often been found as well as salt brine. In Ohio, in the valley of the Muskingum from Zanesville to Marietta, salt wells produced petroleum. Thirty miles north of Marietta, on Duck Creek, a salt well dug in 1814 to a depth of 475 feet discharged "periodically, at intervals of from two to four days and from three to six hours' duration, thirty to sixty gallons of petroleum at each inception."[2]

Dr. S. P. Hildreth, of Marietta, in an account of that region written in 1819, prophetically reported:

> They have sunk two wells more than four hundred feet; one of them affords a strong and pure water, but not in great quantity; the other discharges such vast quantities of petroleum, or as it is vulgarly called "Seneca Oil" . . . that it makes little or no salt. Nevertheless, the petroleum affords considerable profit and is beginning to be in demand for workshops and

[2] S. F. Peckham, "Report on the Production, Technology, and Uses of Petroleum and Its Products," 47 Cong., 2 sess., *House Misc. Doc. 42*, Part 10, pp. 1–154. Mention of oil near Paola, Kansas, appears in *Kansas, A Guide to the Sunflower State*, 88. For early activity in Indian Territory, see the article by Muriel H. Wright, "First Oklahoma Oil Was Produced in 1859," in *Chronicles of Oklahoma*, Vol. IV, No. 4 (December, 1926), 322–28.

manufactories. It affords a clear brisk light, when burned in this way, and will be a valuable article for lighting the street-lamps in the future cities of Ohio.[3]

As early as 1806, oil was found when salt brine was sought near the Kanawha River in western Virginia. David and Joseph Ruffner put down a well near Great Buffalo Lick on Campbell Creek, which they abandoned because it produced petroleum rather than salt water. Other wells of the region likewise produced oil, which was allowed to waste into the Kanawha. "It was from this that the river received the nickname of 'Old Greasy,' by which it was long known to Kanawha boatmen and others."[4]

In 1818, David Beatty bored a well for salt 170 feet deep on the Big South Fork of the Cumberland River, in eastern Kentucky. A flow of petroleum caused its abandonment. This was the first flowing oil well.

A decade later the famous American well was bored near the bed of Little Rennox Creek in Cumberland County, Kentucky. An issue of *Niles' Weekly Register* in 1829 included this account of the 180-foot well:

> Some months since, in the act of boring for salt-water on the land of Mr. Lemuel Stockton, situated in the county of Cumberland, Kentucky, a run of pure oil was struck, from which it is almost incredible what quantities of the substance issued. The discharges were by floods, at intervals from two to five minutes, at each flow vomiting forth many barrels of pure oil. I witnessed myself, on a shaft that stood upright by the aperture in the rock from which it issued, marks of oil 25 to 30 feet perpendicularly above the rock. These floods continued for three or four weeks, when they subsided to a constant stream, affording many thousands of gallons per day. This well is between a quarter and a half-mile from the banks of

[3] Quoted in Peckham, *op. cit.,* 7.
[4] *Ibid.*

5

the Cumberland River, on a small rill (creek) down which it runs into the Cumberland. It was traced as far down the Cumberland as Gallatin, in Sumner County, Tennessee, nearly a hundred miles. For many miles it covered the whole surface of the river and its marks are now found on the rocks on each bank.

About two miles below the point on which it touched the river, it was set on fire by a boy, and the effect was grand beyond description. An old gentleman who witnessed it says he has seen several cities on fire, but that he never beheld anything like the flames which rose from the bosom of the Cumberland to touch the very clouds.[5]

In 1848, after Samuel Kier had popularized the use for medicinal purposes of oil produced from salt wells near Tarentum, Pennsylvania, Job Moses, a resident of Buffalo, New York, bought the property on Rennox Creek. The well was deepened to 400 feet and produced five or six barrels of petroleum per day. This he put up in half-pint bottles for sale at fifty cents as "American Rock Oil," commending it as a specific for numerous body ailments.[6]

During the 1830's, salt wells were sunk to a depth of four to five hundred feet near Tarentum, twenty miles up the Allegheny River from Pittsburgh. Large quantities of salt were produced from the brine and shipped to Pittsburgh. Sometime in the mid-forties, several years after salt manufacture had been underway, certain of the wells also began to produce oil. A resident recalled:

It was in 1844 or 1845 before petroleum appeared in the product of the Peterson salt well. I was working down at the salt works then, driving and doing general work, and remained there for several years. The dirty grease that thus came up

[5] *Ibid.*, 8.

[6] John J. McLaurin, *Sketches in Crude Oil*, 38–39.

from the well with the salt water annoyed us for a time very much. We pumped the water into a tank, and there the oil would gather on the top, enabling us to draw off the water underneath. But after awhile—in the winter, I think—some of the petroleum or "rock oil" as we only knew it then, got into the vats and threatened to injure our apparatus as well as the finished salt, and Mr. Peterson gave orders to run the tank over the next time so that the oil could not possibly get into the vats. We did this, and the oil, which in that way flowed over the sides of the tank, ran into the canal.

Another old-timer recalled:

The oil was wasted for a long time, and when the Kier salt well, a year after appearance of oil in the Peterson well, also began to produce oil, the quantity of grease flowing onto the canal was quite large. The boatmen got to complaining about it. I have heard them myself. They said it greased their tow-lines, making the ropes hard to handle, besides soiling the sides and decks of their canal boats.[7]

The Kier well produced about one barrel of petroleum a day and the Peterson well, two barrels. Samuel M. Kier of Pittsburgh, son of Thomas Kier, on whose place was located one of the producing wells, thought petroleum should have a financial and commercial value. He was already a successful businessman engaged in such varied enterprises as brick plants, a pottery works, the coal mining industry, an iron furnace and foundry, and a canal-boat transportation company and was assisting his father financially with the brine wells.

His earliest marketing attempts have been described in the reminiscences of a former employee:

Wasn't it in 1849 that the gold fever broke out in California? Yes, I thought it was, for I remember that Mr. Kier

7 "S. M. Kier, the Pioneer Oil Refiner," *Pennsylvania Petroleum,* 13–18.

sent dozens of his little boxes of "petroleum butter" away to California in that year. "Petroleum butter" was the buttery-like stuff that oozed up from our salt well on Kier's place with each stroke of the sucker rod. It was formed by the petroleum getting in the iron parts and churning from the friction of the machinery. Mr. Kier had us scrape this off the iron each day and send it to his place in Pittsburgh. There it was put up in little boxes and sold for burns, scalds and bruises—and I tell you it was good for them, too.

The "rock oil" which he sold in bottles for medicine was simply the crude oil of today, though there is no question that that found in the Kier well was of the very best. I have taken many a dose of it inwardly, and, sir, if you ever get a bad cold in the chest, there is no better remedy today than to soak a flannel cloth with crude petroleum and lay it across your breast. Try it some night. In those days everybody up here in Tarentum used the Kier oil for medicine, and I'll bet you will find plenty of persons still living here who yet believe in the virtues of petroleum as a medicine. I am never without half a barrel of crude oil now in the house, and it is my stand-ard remedy.[8]

Kier believed in the curative properties of petroleum as a panacea for many human ailments. He set up a sales organization in medicine-show style that traveled the countryside offering "Kier's Petroleum or Rock Oil" in half-pint bottles for fifty cents a bottle. In 1852, he abandoned wagon sales and placed the product in many drugstores throughout the eastern states.

Two years earlier, Kier had begun experiments in the im-provement of crude oil as an illuminant. His first experiments were done in Pittsburgh with a one-barrel still. He simply heated the oil until it vaporized, then condensed it by cooling. Next, he devised a burner to fit lamps then in use for burning camphine

8 *Ibid.*, 17.

oil, a turpentine product. Local demand created the chief market for his distilled oil.[9]

The Pittsburgh firm of Mackeown and Finley also became interested in marketing crude oil. A. C. Ferris, a New York businessman, visited Pittsburgh in November, 1857, where a druggist showed him a sample of oil. He had a five-gallon can of it shipped to New York and experimented with it by burning it in various kinds of lamps. He decided the product could be marketed successfully and in February, 1858, contracted with Mackeown and Finley for two-thirds of their manufacture of carbon oil. His first sales were made that year in very small lots of from half a gallon to one gallon. Ferris employed salesmen in Brooklyn who, "with a can of oil in one hand and a small lamp in the other, visited grocery and drug stores, thinking they had achieved success on persuading a dealer to order five gallons, with lamps to match, with the condition that the goods were not to be paid for until sold, and to be returned if not sold within a given time."[10] By the end of the year, a profitable business had been built up in the New York area.

In 1856, the firm of Peterson and Irwin bought a farm near Tarentum on which was located a producing salt well. This well produced two and one-half to five barrels of oil a day, which was marketed in Baltimore, Maryland, by Mackeown, Nevin and Company of Pittsburgh. Its principal use was in oiling the wool processed at carding mills in Baltimore. There is a tradition, too, that small quantities of crude oil shipped to England were bottled there with other ingredients and returned to America for sale in drugstores as British Oil.

The success of Kier's Rock Oil and the marketing of carbon oil for more than local use prompted competition. Brewer, Watson, and Company, a lumber firm, was cutting timber along Oil

[9] "Samuel Kier, First Refiner, Was Hot after Petroleum before Easterners Arrived," in the Titusville (Pennsylvania) *Herald,* August 22, 1934, p. 31.

[10] "A. C. Ferris Introduces Petroleum in New York City," *Pennsylvania Petroleum,* 37.

Creek. Its holdings included an oil spring below Titusville near their upper mill site. In the neighborhood were hundreds of cribbed pits, many years old, evidently constructed by Indians to hold oil seepage long before white settlement. Ebenezer Brewer and James Rynd, members of the firm who resided in Pittsburgh, knew of Kier's product and interested other members of the company in an attempt to capture some of the oil seepage.

Brewer, in 1849, sent five gallons of oil from the spring to his son, F. B. Brewer, who was practicing medicine in Vermont. Dr. Brewer gave some of it to Dr. Dixi Crosby, dean of the New Hampshire Medical School, and to O. P. Hubbard, professor of chemistry at Dartmouth College. Both gentlemen considered that the oil had possibilities for medicinal purposes. A son of Dr. Crosby, Albert, foresaw its use for other purposes.

Three years later Dr. Brewer moved to Titusville and became a member of the lumber firm. On-the-spot visitations to springs and seepages carrying oil convinced him that properties owned by the company should be exploited. In July, 1853, the oil spring near the Upper Mill was leased to a near-by resident, J. D. Angier, who collected the oil for one-half interest. Where seepage was indicated, Angier dug ditches or trenches leading to a central basin and skimmed oil from the top of the water.[11]

The following year Dr. Brewer visited New England and again met Albert Crosby, who had his enthusiasm kindled by the doctor's stories relative to the abundance of surface oil indications. Crosby accompanied Brewer on his return to Titusville. They made trips down Oil Creek. These visits convinced Crosby that a profitable promotional scheme should be effected to exploit the natural phenomena.

Crosby entered into an agreement at Titusville with Brewer, representing the lumber firm, for the transfer of 100 acres of land which included the spring near the Upper Mill being worked by Angier. This agreement provided for the payment to the

[11] "Dr. F. B. Brewer's Story," in *ibid.*, 45–50.

lumber company of $5,000 for the 100 acres and oil rights to other lands controlled by the firm, including several hundred acres along Oil Creek and its tributaries. Crosby agreed to raise sufficient capital to develop the oil properties. He proposed to organize a stock company, capitalized at $250,000. The firm of Brewer, Watson and Company was to receive one-fifth of the stock.

Crosby, more interested in the promotional scheme than in the actual development of the properties, was unable to interest eastern capital in the venture. Among those who had become interested, however, were two young New Yorkers, George H. Bissell and J. G. Eveleth, who recently had established a partnership in the practice of law. Bissell, a Dartmouth graduate, on a visit to the college campus had seen a sample of petroleum in the office of Professor Crosby. He agreed with the professor that it probably possessed great commercial possibilities. He and his partner visited Titusville, saw possibilities for development of an oil business, and made an agreement with the lumber firm essentially similar to the abortive one of Albert Crosby. They agreed to pay $5,000 for 100 acres, which included the oil spring near the mill site, and to lease an adjoining 112 acres.

Bissell and Eveleth organized the Pennsylvania Rock Oil Company of New York on December 30, 1854.[12] They were chosen directors, as were Franklin Reed, James H. Salisbury, and Dexter A. Hawkins of New York, Anson Sheldon, New Haven, and Dr. Francis B. Brewer, Titusville. The company was capitalized at $250,000.

The directors were advised that, in order to promote the sale of stock, a prospectus should be prepared which would include an analysis of oil and point up its economic value. Bissell and Eveleth visited New Haven and contracted with Benjamin Silliman, Jr., professor of chemistry at Yale University, to make the

[12] "Articles of Association of the Pennsylvania Rock Oil Company of New York," in *ibid.*, 125–26.

11

analysis. They furnished funds to Silliman for the purchase of necessary apparatus to be used in conducting the experiments and, also, underwrote his fee.

Silliman spent four months on his experiments. His report, published in May, 1855, indicated that petroleum could produce a better, more economical illuminant than any marketable product.[13] The analysis also stressed potential lubricating qualities: "As this oil does not gum or become acid by exposure, it possesses in that, as well as in its wonderful resistance to extreme cold, important qualities for a lubricator." Silliman added: "In conclusion, gentlemen, it appears to me that there is much ground for encouragement in the belief that your Company have in their possession a raw material from which, by simple and not expensive process they may manufacture very valuable products."

The Pennsylvania Rock Oil Company was converted to a stock company under the laws of Connecticut in September.[14] This was done because a statute of New York held the property of stockholders of a company organized in that state liable for the debts of the company. Also, businessmen of New Haven had their interest aroused by the Silliman report. Headquarters for the company were established at New Haven. Twelve thousand shares of stock were authorized in the amount of $300,000. Silliman was elected the first president of the company. In addition to leading subscribers from the New York company, James M.

[13] "Silliman's Report," in *ibid.*, 127–29.

[14] Anson Sheldon to Dr. F. B. Brewer, September 25, 1855, in *The Beginnings of the Petroleum Industry, Sources and Bibliography,* by Paul H. Giddens. The sources contain fifty-three letters to or from Dr. Francis Beattie Brewer; all but one were written between September 11, 1854, and December 12, 1855, and pertain to the promotion of oil development that led to the Drake well. Also, nineteen letters to or from George H. Bissell, written between March 1, 1855, and November 27, 1859, are included in the collection. To these sources are added an invaluable bibliography on the beginnings of the Pennsylvania petroleum industry, compiled by Paul H. Giddens as the result of several years' research.

Townsend, Asahel Pierpont, and W. A. Ives of New Haven subscribed to the enterprise.

Sometime in the summer of 1856, Bissell, while in a drugstore in New York City, picked up and examined a circular advertisement for "Kier's Rock Oil." The circular showed two derricks in Pennsylvania, and described how rock oil was recovered in boring for salt water four hundred feet below the earth's surface and pumped up with the brine. This prompted Bissell to consider the possibility of drilling for oil rather than relying upon sources of seepage on his company's property.

Directors of the Pennsylvania Rock Oil Company agreed it was high time for the development of the property, but New Haven and New York directors experienced a conflict of interests in reaching a decision. At length, on November 26, 1856, a lease was executed to David H. Lyman and Rensseland N. Havens of New York, which empowered them to "bore, dig, mine, search for and obtain in and from such lands, oil, salt water, coal and all other materials and minerals" for a period of ten years beginning January 1, 1857.[15] The lease provided that Lyman and Havens should bear all expense of development and pay to the company "in lieu of all other claims or charges for said lease by way of rent or otherwise, on all the oil produced by them from such lands not exceeding in amount an average of five hundred (500) gallons for each working day in any one quarter year twenty-two (22) cents per gallon, on all oil so procured by them exceeding such average of five hundred (500) gallons, and being less than an average of one thousand (1,000) gallons estimated in the same manner, twenty (20) cents per gallon; and on all oil so procured by them exceeding such average of one thousand (1,000) gallons, fifteen (15) cents per gallon."

[15] A copy of this lease was reproduced in the "Diamond Jubilee of Oil Edition" of the Titusville *Herald,* August 22, 1934.

Lyman and Havens, real estate operators, foresaw this adventure in western development as a means of adding to their company's influence and importance; they could not foresee that overspeculation in the future of the country, particularly overinvestment in railways, was to bring a sudden and great, although temporary, panic. An editorial in *Harper's Weekly* summed up conditions faced by the country in the fall of 1857: "It is a gloomy moment in history. Not for many years has there been so much grave and deep apprehension. . . . In our country there is universal commercial prostration and panic; and thousands of our poorest fellow citizens are turned out against the approaching winter without employment and without prospect of it." Before the real estate company could make serious plans for development of the oil property, the firm became a victim of the depression. They got out of the lease agreement when it was discovered that the wives of Eben Brewer and James Rynd of Pittsburgh had not signed the original conveyances made by Brewer, Watson and Company.

James M. Townsend of New Haven, one of the influential stockholders of the Pennsylvania Rock Oil Company, assumed a more active interest in its organization.[16] The company had five directors and the bylaws provided that the president, secretary and treasurer, and three directors should reside in New Haven County. The three were James M. Townsend, William A. Ives, and Asahel Pierpont. The other two were George H. Bissell of New York and Jonathan Watson of Titusville. Led by Townsend, the New Haven representatives on the board of directors decided that, with the Lyman-Havens lease in default, they should assume responsibility for development of the property. In order to quiet title to the lands obtained from the lumber firm, however, it was necessary to obtain the missing signatures from the original conveyance. Townsend selected Edwin L. Drake, of New Haven, for this purpose.[17] He was directed to

[16] "James M. Townsend's Story," in *Pennsylvania Petroleum*, 53–61.

14

go to Pennsylvania, obtain the missing signatures, visit and examine the holdings of the company, and make a report on his findings.

The man whom Townsend trusted with this mission had no particular talents for the task. He was not a lawyer versed in searching titles for irregularities; he was not even a real estate operator with an understanding of abstracts; he had no knowledge of petroleum, of salt wells, or of salt manufacture; he was not a successful businessman; his years of productive employment had been spent in the service of others, principally in transportation.

Drake was born on March 29, 1819, at Greenville, New York. Six years later his parents moved to Castleton, Vermont, where he received a common school education. After age nineteen he found employment on an Erie Canal boat, then as clerk for a steamship line operating between Buffalo and New York. Next, Drake spent a year on an uncle's farm near Ann Arbor, Michigan, then two years in the neighboring community of Tecumseh as hotel clerk and, afterwards, found employment in a textile factory. From 1842 to 1844, he worked as a salesman in New Haven and New York. For five years he was an express messenger on the Boston and Albany Railroad, and then from 1849 to 1857 he was a conductor for the New York and New Haven Railroad.

Townsend and Drake resided at the Tontine Hotel. The young banker liked the older man. He knew that the railroad man, although extremely reticent about business affairs, was a teller of droll stories and a good companion among men, with a reputation for integrity, honesty, and fair dealing. The banker trusted him.

Drake was available to make the trip west. He was recuperating from a severe spell of sickness in December and had not returned to his regular employment. He possessed, too, a rail-

[17] Titusville *Herald*, August 22, 1934.

road pass, which he could use to Erie, within forty-five miles of Titusville. This he did. The trip from Erie to his inland destination was later described by him, as follows: "At that time there was no railroad to Titusville from Erie and staging anything but agreeable, but as there was no other way to get there, unless I walked, I took my seat in the stage one bright Monday morning for a ride of forty-five miles. The road for the first fifteen miles was very good, but from Waterford, to Titusville there did not seem to be much road, rather an extensive mud hole."[18] He described how the team plodded slowly through the mud for two days and finally arrived at Titusville at ten o'clock Tuesday night.

It was possibly at this time that the erstwhile railroad conductor had thrust upon him the honorary title of "Colonel." Legend has it that Townsend, in order to add prestige to the development scheme, had addressed mail to "Colonel Edwin L. Drake," which awaited his arrival in Titusville. The honorary title stuck. Thereafter, so long as he lived, and since, he was and has been referred to as "Colonel" Drake.

Drake obtained the missing signatures to the original conveyance, made a personal inspection of the company's holdings below Titusville, returned to New Haven, and reported to Townsend and the other directors on his findings. New Haven directors of the Pennsylvania Rock Oil Company then drew up a lease agreement, on December 30, 1857, with Drake and E. B. Bowditch, both of whom were stockholders, to develop the Pennsylvania property.

The lease substantially followed the one made with Lyman and Havens on November 26, 1856. Drake and Bowditch agreed to pay one-eighth of all oil procured to the Pennsylvania Rock Oil Company, which was to furnish barrels for its share. The lessees, too, were given an option to purchase the one-eighth

[18] "Colonel E. L. Drake's Story, Written in 1870," *Pennsylvania Petroleum,* 65.

16

royalty at the source, for forty-five cents a gallon. The company agreed to pay thirty-three cents a gallon on oil that could not be sold in excess of 1,000 gallons each working day. The lease was to extend through fifteen years. By mutual consent the lease was modified on February 12, to provide for renewal for twenty-five years and for the lessees to pay a royalty of twelve cents a gallon on production in lieu of one-eighth part.

Townsend and other New Haven representatives of the parent company then proceeded to organize the Seneca Oil Company, a joint-stock corporation, under the laws of Connecticut. The purposes of the corporation were stated in the articles of incorporation: "To raise, procure, manufacture and sell oils of all kinds, paints, salt, coal or any mineral productions which may be found in any springs or mines on any lands that may come into possession of said company by deed or lease."[19] Drake and Bowditch transferred their lease to the company. At the first meeting of the stockholders, on March 19, Drake was appointed chairman as well as one of the three directors; the others selected were W. A. Ives and J. F. Marchall. On March 23, the number of directors was increased to seven by the addition of Asahel Pierpont, Edwin B. Bowditch, James M. Townsend, and Henry L. Pierpont.

At the next meeting of the Board of Directors of the Seneca Oil Company, on April 1, 1858, it was voted that "Edwin L. Drake be appointed general agent of this Company to raise and dispose of Oil with a Salary of one Thousand Dollars per Annum from the date hereof." The directors agreed, too, that an additional sum of one thousand dollars should be made available to Drake for operating expenses. He was authorized to proceed to Titusville as soon as possible and get work underway.

[19] "Organization of the Seneca Oil Company," in *ibid.*, 139–40.

2

DRAKE'S FOLLY

DRAKE, as nominal president and general agent of the Seneca Oil Company, returned to Titusville in May, 1858. He intended to excavate the principal oil spring located on an island formed by the millrace and creek approximately two hundred feet south of the lumber mill.

The spot selected was in a tight valley hemmed by hills two to three hundred feet in height, rising almost from the narrow creek bed—hills that had been covered with the virgin maple, walnut, chestnut, oak, pine, beech, and hickory forests, and with smaller trees and shrubs: dogwood, ironwood, willows near the water's edge, mountain laurel, and juneberry.

Already the lumber company had slashed acre after acre of the valuable timberland. Drake, later, wrote this description of Titusville and vicinity:

> . . . on Tuesday night I arrived at the now famous City of Titusville, but at that time a small delapidated Village containing about 125 souls of all descriptions. The place had been settled about sixty years, and had been and was at that time, the central point or headquarters of a heavy lumber company (Brewer, Watson and Company) which had been engaged several years in making as well as buying of all the small dealers in that vicinity all the lumber they could, and running it to Pittsburgh and markets below there, and selling it. Although they converted the lumber into cash but very little of

18

that cash ever found its way into the pockets of the inhabitants of Titusville and vicinity (except the members of the firm) as one man told me he had worked for them fifteen years steady at one dollar per day and only received in that time twenty-one dollars and fifty cents in cash; the balance of his earnings he had received in goods from the store of the company and those at such high prices that he was then in debt to the company nearly three hundred dollars. I merely mention this particular case as an illustration, showing that although the lumber company was becoming rich, the inhabitants were becoming poorer every year and as one of the firm (Dr. F. B. Brewer) told me at that time that he knew just how many pine trees the company owned and about how much lumber they would make, and with their present facilities for sawing, it would take them about five years to cut up the logs and run the lumber to Pittsburgh, and, said he, we shall leave here and the place will die a natural death. There will be no business. The people will have to starve or move, and in ten years the main street would be grown up to grass, and I must say it really looked as if it might come to that as the grass grew in some of the streets at that time.[1]

A rainy season and high waters from Oil Creek delayed operations, although Drake continued skimming operations already in effect, adding shallow pits and interlocking ditches wherever oil seepage appeared. The digging of a hole approximately ten feet square at the principal spring was slow, tedious work, delayed by flooding and handicapped because the bottom of the excavation was below the level of Oil Creek. Percolating water continuously filled the hole. Power from the mill was used in pumping water from the excavation, but it was rarely available when the workmen needed it.

Drake reported to Seneca officials on July 2:

[1] "Colonel E. L. Drake's Story, Written in 1870," *Pennsylvania Petroleum*, 65.

Here I am digging along yet in search of oil and other valuables. The month of May was a hard one, and the first eleven days of June, but since then we have had dry weather, so that I have got the start of the water, and am now gathering about ten gallons of oil per day—at the same time sinking a well for the purpose of taking what oil there is on the island.

I have found some difficulty in getting a borer. All were engaged on jobs that will last until fall. Yesterday Dr. Brewer wrote me he could get one for me at Allegany, who will bore and tube for three dollars per foot, which is the best offer I have had. I wrote the Doctor to send him along at once. Yesterday I set some men to opening a new spring, so that things begin to look greasy.[2]

The phrase "sinking a well," used in the letter, was a common expression in vogue among well diggers seeking veins of fresh or salt water. It was the universal practice to dig a hole large enough for workmen to wield ax and shovel through soil and hardpan to bedrock, and there begin the drilling operation. Drake's reference was to the ten-foot-square excavation that was to be dug, then cribbed as protection against cave-ins and surface or ground water, before actual drilling began.

On his first trip to Titusville, Drake had stopped at Syracuse, New York, and observed drilling operations for a salt well. Recently he had seen similar operations near Tarentum, Pennsylvania. It was natural, therefore, that his well-drilling attempt should follow time-honored practices.

As the summer advanced, however, Drake became increasingly impatient, because of the flooding conditions at the excavation, pump breakdowns, and the limited amount of money furnished by the Seneca Company for operating expenses.

Some of these perplexities were expressed in a letter written August 16:

2 E. L. Drake to James M. Townsend, in *ibid.*, 144.

In sinking our well last week we struck a large vein of oil but the same thrust of the spade opened a vein of water that drove the men out of the well and I shall not try to dig by hand anymore as I am satisfied that boring is cheapest. I should have had my borer here but I wrote him on the first I was not ready as I did not know that you could raise the money. But money we must have if we make anything. I have abandoned the idea of boring and pumping by water as I could not have the exclusive use of the power, but must be subject to the sawyer, the turner, and the blacksmith; so after consulting the best businessmen—that is, salt and oil men—at Tarentum, I have contracted for an engine to be ready for boring by the first of September. The engine will cost five hundred dollars in Erie which is about one hundred dollars less than the same or one like it would cost at the East.

When I get that I shall be independent of the lumber company. I have had my pump stopped ten days this month and, at the rate they repair, it will take all the month of September to repair the flumes and wheels, and then I am liable to be stopped at any time, which will not answer.

I have got out the timber for my pump house and am having it framed today. We shall get that up this week and then I want to get out timber for a building 60 feet long and 30 feet wide for oil vats and salt pans which I intend to put up in September, and to do this will require some money.

Now I think you had better make a loan of $1,000.00 and place it in bank there where I can get it, as I need it and I assure you there is no risk whatever for I have got as far with five hundred dollars as any other company here with five thousand, and further than some have with ten thousand dollars.

I shall send in a statement of my stewardship the first of September and, in the meantime, if the Seneca Oil Company should feel as if they were too poor to furnish $1,000.00 more by the tenth of September, please let me know at once. Money is very scarce here. The lumbermen could not sell their lumber for cash this summer and the people all depend upon the lumber trade, so money is as tight here as it was in New York last fall. The old lumber company begin to think they did not retain the best of the property when they sold out the oil springs. Old Mr. Brewer is here now and says he is sorry they sold that piece of land or gave that lease; but let them whine; there is more money in that little island than there is in all the 1200 acres of the lumber company's land.[3]

Drake had received $500 in cash from the Seneca Oil Company on April 20, and an additional $500 was forwarded to him in August. He used the funds for on-site operations of the company, necessary travel on company business, tools, equipment, labor charges, and state and county taxes. Principal expenses incurred during this season of frustration were for the pump, for labor attendant to the skimming operations underway, for cleaning and repairing ditches, for the construction of ditches to additional sources of oil seepage, and for the digging of the well at the Oil Spring site.

Drake's "folly" attracted attention and comment far beyond the valley of Oil Creek. He had trouble in his attempt to hire men "to work for a lunatic." Finally, at Tarentum, one hundred miles away, a well digger promised to come to Titusville and bore a smooth, round five-inch hole, one thousand feet deep, for $1.25 a foot. He was to be paid at the completion of the contract and, while the work was in progress, draw only enough pay for board and tobacco.

When the man had not arrived to begin work early in August, Drake again rode to Tarentum. The salt driller promised to begin drilling in September. Again, Drake was disappointed. Again, he made the arduous trip by horseback to Tarentum, where he accidentally learned why the driller had disappointed him. He regarded Drake as crazy and "thought the easiest way to get rid of him was to make a contract and pretend that he meant to come."

[3] E. L. Drake to W. A. Ives, in *ibid.*, 144–46.

Drake then rode to Pomroy, Ohio, to Pittsburgh, and back to Tarentum seeking a driller. By this time it was mid-November. A friend at Tarentum, Lewis B. Peterson, a salt manufacturer, advised him to wait until spring before resuming plans for drilling.

Drake accepted the advice offered by his friend. He returned to Titusville and supervised the building of an enginehouse at the site for drilling operations. In October, he was advised by the company not to do anything about the manufacturing of salt unless the brine was unusually strong and could be inexpensively processed. It was suggested he might prospect for coal "if you have spare time when it will not interfere with the main object."[4]

The lease agreement under which Drake was operating included identical phraseology of the defunct Lyman-Havens agreement: To "bore, dig, mine, search for and obtain in and from such lands, oil, salt water, coal and all other materials and minerals. . . ." The primary aim of the company, of course, was to obtain oil, but other mineral properties were not to be overlooked. Drake devoted his energy to the main object of the company—oil.

The company forwarded $505 to Drake on October 30, $483 on December 30, and an additional $500 on April 2 of the following year. He purchased an engine and boiler, paid for lumber and carpentry, team hire and labor. After the enginehouse was completed, severe winter weather forced postponement of further activity.

His friend Peterson, of Tarentum, wrote him in April recommending a well driller whom he had employed on salt well projects during the past several years. Drake immediately went to Tarentum and met the driller, William A. Smith, who promised to come to Titusville after finishing jobs on hand that would take about one month to complete. He was to receive $2.50 a

---

[4] James M. Townsend to E. L. Drake, in *ibid.*, 147–48.

day. This wage also included the services of his sixteen-year-old son, Samuel B. Smith.

Smith was a blacksmith and well driller who lived in Salina, about one and one-half miles from Tarentum. He was sober and industrious and had a good reputation. Drake satisfied himself on these qualifications. Later, he wrote:

> For I had learned from an old Salt Man (or Salt Manufacturer rather) that many of the borers were thirsty souls and preferred Whiskey to any other liquid for a steady drink, and not infrequently a hole four or five hundred feet deep had been spoiled by the unskilful management of a drunken Borer, thereby causing a loss to the owner of not only the amount paid for boring but also the time necessary for boring another well. I was determined not to pay for any man's carelessness.[5]

Smith had had experience in manufacturing drilling tools and in boring salt wells. Drake bought iron and steel for him, and he forged bits, temper screws, jars, and sinker bars in his shop. In mid-May, Drake hired a team and wagon to transport the drilling equipment, which cost $76.50, Smith, his son Samuel, and grown daughter Margaret to Titusville.

The family group converted the enginehouse, constructed at the well site the previous fall, into living quarters by subdividing it into a bedroom and kitchen; later, a lean-to was added. As soon as the temporary home was fixed, the Smiths constructed a shed to house the engine and boiler.

Drake, meantime, had employed Samuel Silliman, a Titusville carpenter, to frame big, square timbers for a derrick. Years later, Samuel Smith described how it was placed in position:

> I thought that rig raising one of the most interesting of all the things connected with the drilling of the first well. The raising was done by volunteers much like the old-time barn

[5] "Colonel E. L. Drake's Story," *ibid.*, 66–67.

raisings, known to every farmer in the United States. I guess 20 or 30 men must have been there at the time. Probably Colonel Drake did not need that number of men, but the men were there just the same. Some came from the Upper Mill, some from the Lower Mill, and some from Titusville. For though Titusville people did not quite understand Colonel Drake and his oil well, they were all greatly interested in what was going on. And whenever they could be of any help, as in this matter of the derrick raising, they were not only willing and anxious but even eager to help.

It must have taken at least an hour to raise that derrick. The boards were not nailed on by the "crowd" during the afternoon of the derrick raising. They were placed in position the next day. And when they were finally nailed on, the derrick was not boarded all the way to the top.[6]

The dug well, which was commenced in the summer of 1858, was about sixteen feet deep by the time Smith had the boiler and engine housed and the derrick in place. Because the hole was below the depth of the surface of Oil Creek, however, water percolated through the porous soil so rapidly that pumping operations were unable to draw it off in sufficient volume to permit much accomplishment on the extension of the hole. Workmen employed in the drudgery of digging found, too, that the water was so cold they could work no more than ten minutes at a time.

Drake considered this problem. His solution, simple though it proved to be, presaged a revolutionary step in drilling methods.[7] He decided that an iron tube should be driven to bedrock, and that this tubing would shield drilling operations from cave-ins and, where the soil was porous, such as that at the Oil Spring,

[6] "Recollections of Samuel B. Smith and James P. Smith," *ibid.*, 83.

[7] E. L. Drake to J. M. McCarthy, June 28, 1872, in *List of Letters Belonging to Dr. Francis Beattie Brewer,* published in Giddens' *The Beginnings of the Petroleum Industry,* 59, Drake stated: "I claim that I did invent the driving pipe and drive it and without that they could not bore on bottom lands when the earth is full of water." This and, of course, his determination and perseverance in completion of the first oil well were his chief contributions to the oil industry.

from percolating waters. He drove to Erie, Pennsylvania, visited the firm of Liddell and Marsh, and explained the design and type of cast-iron tubing he wanted from the manufacturing concern. He ordered fifty feet of driving pipe cut in ten-foot sections, of three-inch bore and one and one-half inches thick.

When the driving pipe was delivered at the well site, Drake found it was only one-half-inch thick. He attempted to use this, but when the second section was driven, it caused the first section to break near its top. Again, he experienced delay. He returned to Erie and this time obtained pipe constructed according to his specifications.

Directors of the Seneca Oil Company had lost faith in their venture because of the continual drain of capital and lack of return on their investment. During the period from April, 1858, to September, 1859, Drake received $2,490 from the company to finance exploratory activities. James M. Townsend, New Haven, was the principal source of this financial backing. Some years later he wrote: "The raising of the money and sending it out was done by me. I do not say it egotistically but only as a matter of truth, that if I had not done what I did in favor of developing Petroleum it would not have been developed at that time."[8] Although he called upon prospective investors in New Haven and New York to interest them in financial backing, they would listen, look at him, and shake their heads with a remark something like this: "Oh, Townsend! Oil coming out of the ground, pumping oil out of the earth as you pump water? Nonsense. You are crazy." Finally, in August, 1859, after other stockholders and directors of the company either refused or were unable to advance additional funds for the experimental project, Townsend decided he was backing a losing venture. Near the end of the month he sent a final remittance to Drake and advised him to pay all bills and return to New Haven.

Additional financial assistance was obtained by Drake from

[8] "James M. Townsend's Story," *Pennsylvania Petroleum*, 60.

other sources. Months spent in the small village of Titusville had won friends who admired his courage and persistence. Reuel D. Fletcher, a young merchant, had extended him credit at his store since June 5, 1858, a few weeks after Drake, his wife, son, and infant daughter had moved to Titusville.[9] He lent Drake the use of a saddle horse for trips to Tarentum, Pittsburgh, and other points. Peter Wilson, the druggist, also permitted him to open an account. Both men liked Drake. A bank had not been established in Titusville, but during the summer of 1859 Drake visited the Dick bank in Meadville, Pennsylvania, to arrange for a loan in the amount of five hundred dollars. Fletcher and Wilson endorsed the note for him.

By mid-August, the driving pipe made to order by Liddell and Marsh had been delivered to the well site. Smith connected sections by bands. A sleeve was fitted loosely over the joint, a hole drilled through the band, and a rivet driven to prevent the band from slipping. The pipe was driven into the ground through the use of a battering-ram, lifted by a windlass which had been used earlier to remove dirt from the hand-dug excavation. The battering-ram—a thick, heavy, eight-foot length of white oak—quickly and effectively drove the pipe approximately thirty-one feet to bedrock.

Smith, at last, was ready to begin drilling operations. A cable, connected to the engine, was hooked to the drilling tools, and a stem, 25 feet long and about 1¾ inches thick, was joined to the steel-lined bit joint. Work progressed slowly and without incident. In order to pump sediment from the well, a copper sand pump was used. Drilling tools were brought out of the hole and unscrewed from the cable socket. A thread on the upper end of the sand pump was screwed into the socket and then lowered by means of engine power. This method of cleaning out the hole was time consuming.

Two bits were alternately used in drilling—one, seven-

---

9 "Reuel D. Fletcher's Story," *ibid.*, 103–16.

eighths of an inch across, the other, one and one-quarter inches. Drake took them on one trip to Warren, Pennsylvania, to the blacksmith shop of John Gilfillen, who sharpened them.

Late Saturday afternoon, on August 27, when drilling at a depth of approximately sixty-nine feet, Smith noticed that the tools dropped a few inches. The bit evidently had penetrated a crevice. The tools were drawn, to remain idle over Sunday, a customary observance.

On Sunday afternoon, Smith walked to the well. He peered down. Some feet below the surface he could see a dark fluid. He plugged one end of a tin water spout, tied it to a string, and let it down into the liquid. He drew it up filled with petroleum. Oil had been struck!

3

DEVELOPMENTS ALONG OIL CREEK

JOHN J. MCLAURIN has described reaction to the Drake discovery:

> The news spread like a Dakota cyclone. Villagers and country-folk flocked to the wonderful well. Smith wrote to Peterson, his former employer: "Come quick, there's oceans of oil!" Jonathan Watson jumped on a horse and galloped down the creek to lease the McClintock farm, where Nathaniel Carey had dipped oil and a timbered crib had been constructed. Henry Potter, Titusville, tied up the lands for miles along the stream, hoping to interest New York capital. William Barnsdall secured the farm north of the Willard. George H. Bissell, who had arranged to be posted by telegraph, bought all the Pennsylvania Rock Oil stock he could find and in four days was at the well. He leased farm after farm on Oil Creek and the Allegheny River, regardless of surface indications or the admonitions of meddling wiseacres.[1]

Drake, meantime, as general manager of the Seneca Oil Company, was busy with company business. He commenced pumping the well, using a common iron water pump for the purpose. He then visited Erie, Pennsylvania, and Cleveland, Ohio, in search of copper tubing for the well, and finally obtained it in Philadelphia. After tubing, the well averaged about

[1] McLaurin, *op. cit.*, 62.

twenty barrels of crude oil each day. Drake never operated the well on Sunday, and as a result more water than oil was pumped the following day.

Drake, too, had to solve an immediate problem of storage. Samuel Silliman, the Titusville carpenter, constructed the first vats out of pine boards and timbers obtained from the Upper Mill, and Drake ferreted out empty whiskey barrels in the village. Soon a brand name, "Seneca Oil Co.," appeared on barrels prepared for marketing.

But marketing proved a major problem. Drake visited Cleveland, Chicago, Cincinnati, and Pittsburgh in order to solicit agencies for distribution of the oil. A letter arrived from London, England, in March, 1860, in which the writer offered to set up a distributorship for Great Britain and Europe. James M. Townsend was interested in establishing a brother, William, as general agent in New York; another brother, Captain Charles Hervey Townsend, took a sample of petroleum to Havre, France, and had it analyzed by A. Gelee, a French chemist, who pronounced that it had excellent qualities for illumination, lubrication, and various chemical uses.[2]

On November 14, Drake signed an article of agreement with Samuel M. Kier, of Pittsburgh, in which Kier agreed to purchase one thousand gallons of oil per week from the well, at sixty cents per gallon.[3] At that time the well was producing more than that amount each day, although a disastrous fire shortly before had destroyed the derrick, the enginehouse, and the oil stored aboveground. Part of the enthusiasm attendant to the Drake well is reflected in a letter George Bissell wrote his wife on Friday, November 4:

> We find here an excitement unparalleled. The whole population are crazy almost. Farms that could have been bought

[2] Letters and reports in *Pennsylvania Petroleum,* 154–62, reflect interest in the well.

[3] The original agreement with Kier appears in Document No. 208 in the Drake Museum, Titusville.

for a trifle 4 months ago, now readily command $200 and $300 an acre, and that too when not a drop of oil has ever been discovered on them. So much for the bare hope of their being by any possibility a sub-stratum of oil. Judge of the value assigned by the people to *our lands* where from one well only they are now raising 1200 gallons of pure oil a day. Brewer has just left us and says that Pittsburgh men consider our property worth millions. Last week Ives, Townsend and all the New Haven men were here. Did I tell you that about 10 days ago, one of the workmen went into the store house with a lighted candle and the oil took fire and during the night burnt over $12,000 worth of oil. That loss, however, is a trifle —they can make it up in a week.

Three days later he wrote:

I caught a very severe cold Saturday in wading through the low lands of this region—could hardly speak above a whisper for two days but am now nearly recovered. I found on the inside of a sleeve of my flannel shirt a strip of red flannel which you forgot to cut off—well, I bound this, well saturated with Rock Oil, about my neck and took repeated doses of the oil. I really think it would have resulted badly for me without this remedy. It is positively a specific for throat ailments of such a nature.

... They are now raising from 1000 to 1200 gallons of pure oil a day from *one* spring on our lands. It costs about 1 cent a gallon to raise it and sells readily from 80 cents to a $1.00 a gallon. Drake has been offered $150,000 for his lease and has refused it. Double that sum will be made this year from our lands—probably much more. No California Placer was ever one tenth part so valuable. When the other springs are opened the profit will be millions. I never saw such excitement. The whole western country are thronging here and fabulous prices are offered for lands in the vicinity where there is a prospect of getting oil. We are negotiating with every prospect

31

of success for a tract on Oil Creek several miles below here where there are strong traces of the presence of oil. If we secure them, Eveleth and I are secure of an enormous fortune. . . . Had Eveleth and I held on to our lands in 5 years we should have been the richest men in New York. As it is, we shall do well, very well—under any possible contingency.[4]

Drake, by March, 1860, had been able to contract for the sale of about thirty-five barrels of crude oil each week. About this time, when in Pittsburgh, he met George M. Mowbray, a chemist associated with the wholesale drug firm of Schieffelin Brothers and Company of New York. On March 12, Mowbray, for his firm, agreed to accept at Jersey City all the production from the Drake well in excess of the one thousand gallons per week contracted for by Kier.[5]

Transportation problems, too, plagued Drake. Although railroad connections at Union were only twenty-two miles distant, the road from Titusville was almost impassable. Mail was twelve hours on the road between the points; teamsters charged $1.00 to $1.25 to haul one barrel of oil from the well to the railroad.

These duties—the employment of teamsters, the purchase of barrels, payrolls to meet, and company affairs to attend—kept Drake fully occupied in serving the interests of the Seneca Oil Company. A more enterprising and less conscientious representative possibly would have foresworn loyalty to the company for self-interest and joined Jonathan Watson on his mad dash down Oil Creek to obtain leases. Land was cheap and leases could be obtained for the asking those first few days, with rates of royalty set at the lessor's option.

Then Drake, too, like many of the curious who flocked to his well, did not realize the vast implication of the discovery.

---

[4] Bissell's letters to his wife, November 4 and 7, appear in Giddens, *The Beginnings of the Petroleum Industry*, 80–82.

[5] "Agreement between E. L. Drake and George M. Mowbray for Marketing Petroleum," *Pennsylvania Petroleum*, 163–64.

When he wrote he had "tapped the mine," he did not mean to imply other regions along Oil Creek might not be productive; rather, he implied that, as the responsible representative of the Seneca Oil Company, duties attendant to the production and marketing of oil from the discovery well should occupy his time and attention. Suddenly there had appeared a product more abundant than man had dreamed, in an inaccessible region and with limited known uses. While within the week, representatives of his firm were in the vicinity and, acting within their individual capacities with wise use of imagination and capital, were furthering their interests, Drake, bound by a lifetime of working for others, loyally devoted his talents to company business. Dr. F. B. Brewer, a stockholder in the Pennsylvania Rock Oil Company, who, before the discovery well was drilled, had exchanged oil stock for cigars, later paid this tribute to Drake: "Honest and upright, he risked his all to develop the oil interest in Pennsylvania, but like many another enterprising man, he shook the boughs for others to gather the fruit."[6]

If this were true of Drake—his lack of personal profit from the discovery—it was equally true of the Seneca Oil Company. While plans were a-borning, in the fall of 1859, to set up integrated control of the production, marketing, refining, and distribution of petroleum, an explosion at the well completely destroyed the derrick, enginehouse, and aboveground storage. Recovery from the disaster was delayed by weather conditions, and when operations were resumed in the spring, better production properties had been developed by rival concerns.

The discovery well probably produced no more than three thousand barrels of crude oil before it was abandoned in 1863. Drake well Number Two, about one hundred feet distant, was completed by Uncle Billy Smith in 1860; by September, 1862, both wells were producing less than fifteen barrels per day.[7]

6 "Dr. F. B. Brewer's Story," *ibid.,* 50.
7 E. L. Drake to James M. Townsend, September 4, 1862, in *ibid.,* 180–81.

Increased production by rival concerns forced the price of crude oil downward, and in September, 1860, the Seneca Oil Company found it impossible to make royalty payments to the Pennsylvania Rock Oil Company of twelve cents a gallon. In return for release from the royalty payments, the Seneca Oil Company gave up its title to all the leasehold with the exception of thirty acres. The released portion, about seventy-five acres, was purchased by George Bissell for $50,000, and the parent company was dissolved.

When flowing wells down Oil Creek brought a glut in production the following year, all shallow production, including the property of the Seneca Oil Company, shut down during the fall and winter months. In March, 1864, the property was sold and the company went out of existence.

By this time, Drake had left the oil field region. In addition to the annual salary of $1,000 paid him for managing the Seneca properties, he served as justice of the peace at Titusville in 1861–62, and his fees brought an additional income of $2,500. During the next two years, he acted as oil purchaser for the Schieffelin Brothers organization.

Shortly after Drake and his family came to Titusville in 1858, they moved into a home on a twenty-five-acre plot at the eastern edge of the village. He purchased the property from Jonathan Watson for $2,500, making a down payment of $500. Drake sold the property in 1864, for $12,000; three years later it was estimated to be worth ten times as much.[8]

Drake left for New York City early that year to engage in oil stock and property brokerage. Two years later he was broke. His health, never good after the illness that forced him to give up railroad employment in 1857, was impaired by ex-

---

[8] Titusville *Herald,* August 22, 1934. Edmund Morris, in *Derrick and Drill,* 107, states: "Colonel Drake did not make a fortune, and is rated a poor man financially. Last fall the oil-men talked of making him a handsome testimonial or donation; but the fact that he suffered himself to be a candidate for Assembly, on the Copperhead ticket, prompted them to abandon the idea for the time."

posure during his oil-drilling efforts, and grew steadily worse after the failure of his brokerage business. In 1869, friends in Titusville learned that for three years he had been unemployed, a constant and almost helpless sufferer from a neuralgic affliction of the spine, living in destitute circumstances in a small cottage provided by a friend, and with only meager support for the family from Mrs. Drake's sewing.

A mass meeting was immediately held in Titusville to raise a relief fund for Drake and his family; more than $4,800 was forwarded to them, and Mrs. Drake wisely administered it to relieve the distress of her husband and their children. In 1873, the Pennsylvania legislature granted him a pension of $1,500 a year, in recognition of his oil discovery, "which discovery has greatly stimulated various industries, and has also added directly to the revenues of the commonwealth more than one million dollars since the discovery." This was a lifetime annuity, which continued as long as he or Mrs. Drake lived.

The family moved to Bethlehem, Pennsylvania, after the passage of the legislative act. Until his death on November 8, 1880, Drake was a constant sufferer and a hopeless invalid. Mrs. Drake lived until May 18, 1916.[9]

Better producing wells were developed along Oil Creek soon after the Drake discovery. Watson had raced southward and obtained an option on the Hamilton McClintock farm. On this property was the most celebrated oil spring along the creek; from it a generation before, Nathaniel Carey had filled wooden casks with petroleum and hauled it to Pittsburgh for local use. J. D. Angier contracted with Watson to produce oil from a cribbed well at the spring; by November, pumping operations produced five to eight barrels of oil each day.

William Barnsdall, Boone Meade, and Henry R. Rouse started a second well, named "The Barnsdall," immediately north of the Seneca property, shortly after the Drake discovery.

[9] *Ibid.*

By November, the well was down 80 feet. Three days of pumping yielded three barrels of oil. Drilling was resumed in December. Early in February, William H. Abbott, a merchant of Newton Falls, Ohio, came to Titusville and invested $10,000 in this property. On February 19, at a depth of 112 feet, into a second sand of production, tubing was put into the well, which pumped fifty barrels per day.

The third well drilled in the vicinity, "The Crossley," by the partnership of William Barnsdall, William H. Abbott, and P. T. Witherop operating under the firm name of Crossley, Witherop, and Company, was located south of the Drake well and across the creek. In March, from a depth of 124 feet, it began producing at the rate of seventy-five barrels a day, from the second sand.

Production from the Barnsdall and Crossley wells renewed excitement in oil development in the region. Typical was the news item that appeared in the Jamestown, New York, *Journal,* on March 28, 1860:

> Sloan and Crossley of Titusville on Monday of last week struck on the steep bank of the creek at 124 feet depth, a large mine of oil yielding ever since fifteen quarts per minute. A patent pail of oil and water is given off every 6 seconds, and the proportion of oil is three pints. This statement is from well known and reliable citizens who were on the ground and tallied the yield by the watch. The greatest excitement exists in that region, and fortunes are made in a few minutes by sale or lease of lands. A Mr. H——— bought 300 acres for $30 per acre, and then leased it for $300 per acre, and ¼ of the oil found. Wells are sinking in every direction, and strangers are flocking in from all parts of the country.[10]

[10] Quoted in Thomas A. Gale's pamphlet, *The Wonder of the Nineteenth Century: Rock Oil in Pennsylvania and Elsewhere* (Riceville, Pennsylvania, 1860). This appears in *Pennsylvania Petroleum,* 204–205.

These wells were "kicked down" by the aid of spring poles, as were hundreds later, for shallow production. A visitor to the region made the following observations in September, 1860:

The apparatus for boring is very simple. A derrick is erected, consisting of four timbers from thirty to forty feet, connected with framing ten feet square at the base, and about four or five at the top. Most of these are boarded on the outside, but many are open, except at the bottom. At the top is a pulley over which a stout rope runs, one end of which is attached to the drill and the other to a windlass. The drill consists of a steel edge or point attached to a long iron bar or rod of three inches diameter. This bar is firmly screwed to another, in which there is a flat link or "jar," as it is termed. There are several of these in one drill, in order to afford play for the rigid iron when it strikes the solid rock. The rope attached to the drill is then fastened firmly to the end of a long spring pole. This pole is secured at the outer end, some distance from the derrick. A springing motion is then given to the smaller end of the pole from which the drill hangs, by various expedients. The simplest is by having a strap of rope suspended from it, with a step-piece at the bottom, in which two men each place a foot. By kicking outward or downward a little, the pole comes down, and the natural spring throws it back to the original position, thus moving the drill up and down a short distance. A man stands by the drill, constantly turning it, to vary the side on which it strikes, and to produce a round hole.

Some have a hinged platform connected with the spring-pole, and two men tramp on this all day. As the hole is made, the drill is lowered, and at frequent intervals it is elevated by the derrick entirely out of the opening, and what is called a sand-pump is lowered to bring out the water, earth, etc. The contents of this vessel are scrutinized closely for indications of oil or gas, and if not satisfactory the drill is again resorted to. The depth at which oil is found varies from 30 feet to 400, the average at McClintock's being 150 feet.

Steam-power is rapidly being introduced; all who are pumping oil making use of engines of about five horse power, and a few are drilling by this means. Occasionally horses are used, but at present it appears that almost every man wants to put his foot into it, and jump himself rich. . . . Where pumps are in operation, five inch cast-iron pipe is sunk into the well, and the oil drawn up is conveyed on rude troughs or pipes to vats several hundred feet distant. This is done for safety, as the oil in its natural state is highly inflammable, and several establishments have been burned down by an accidental spark.

Vats for receiving the oil and separating the water are erected according to the yield of the wells; barrels are filled as fast as they can be obtained, and sent off immediately by wagon to Union, thence by the Sunbury and Erie railroad to Erie and the New York and Erie railroad. All the oil yet obtained, with slight exceptions, has been bought by a firm in Jersey City, who have controlled the market. The opening of new wells daily, however, and the increase of the trade, will undoubtedly cause the oil men to seek other channels of communication. Some is now sent by the river to Pittsburgh. Some enterprising citizens of Meadville have formed a company, and are about to erect a refinery in that place. They now have a small one in operation with successful results.

Barrels, barrels, are the great want now, and much oil is lost daily by the scarcity of this article. We should think a good establishment of this kind would pay large profits, as we only saw one in Titusville. The barrels are sold at $2 apiece, and there is already a demand for a thousand a day. The Williams well, owned by Tanner and Watson, runs from one hundred to one hundred and twenty barrels of oil every twenty-four hours. On Sunday work is suspended, and on Mondays the well yields water only. It requires almost a day's pumping to regain the condition of Saturday as to the flow of oil. This well is one hundred and fifty feet deep. It runs sixty-eight barrels itself.

The estimates given for boring, etc. are about as follows: Tools, $75; derrick, $20; digging above the rock, at $1 per foot, 50 feet, $50; boring, $1.50 per foot for 50 feet, and $2.12½ for 100 feet; total for boring, $282.50; pump, $125; total $557.50. This is for a well of 200 feet. But to this must be added the expense of piping, repairs, incidentals, etc. making at least $1,000. Then the cost of the lease of the land, and the risk run of boring perhaps 200 feet, and getting nothing, must be considered. The total expenses of a good well are calculated at $5 per day, or including barrels, freight, etc. to New York, $30; and if 10 barrels are obtained at 40 cents a gallon, the receipts would be $160 in 24 hours.

Doubtless there will be more failures than successes, for so many rush into the thing without counting the cost, with inexperienced workmen and heavy outlays, with no calculation for accidents. Good advice is given by a Titusville paper, that no one should expend money in this enterprise without being able and willing to lose every dollar invested.

The whole thing is so new, and the supply of oil has been so great, that difficulty is experienced in bringing it into use, and time will be required for that purpose. It is claimed that it has the following uses: 1. As a medical agent in applications to burns, bruises, sprains; also, for lung complaints, colds, rheumatism, etc. 2. For greasing wool. 3. For preparing paints. 4. As an illuminator. 5. For generating steam. 6. As a lubricator. If it answers well for only part of these, its utility will be evident.[11]

The spring-pole method of kicking down a well had been in operation for drilling water or salt wells long before engine power came into use, and afterwards, to minimize expense. A deep drilling was made practicable after William Morris at the Kanawha salt borings in 1831 invented a simple tool, a long double link of iron, with jaws that fit closely, which could slide

[11] Morris, *op. cit.*, 34–39.

loosely up and down above the bar connected to the bit and help break the jar of impact.

From salt-boring experience, too, came the use of the sand pump, described by a visitor to Oil Creek in 1862:

> When, after boring for a given time, the men think that so much rock has been loosened as to render it necessary to clear out the hole, the drill is wound up to the top by means of a windlass, and the sand-pump is lowered. This pump is merely an iron tube, with a valve at the bottom opening inwards. When let down into the bore the valve is forced open by coming in contact with the pounded rock, which gets to the inside. As soon as it is raised the contents of the tube pressing on the valve close it, and so imprison themselves within. The sand-pump has to be used, when going through soft rock, many times a day.[12]

Even the reamer, used to round the hole, was borrowed from salt-boring experience. And the trick of packing the tubing, to keep veins of water from strata above percolating below to the oil sand, was borrowed directly from a practice long in vogue among salt borers. The method used during the 1830's in West Virginia has been described, as follows:

> In the manner of bagging the wells, that is, in forming a water-tight joint around the tube to shut off the weaker waters from the stronger below, a simple arrangement, called a "seed-bag," was fallen upon, which proved very effective. . . . This seed-bag is made of buckskin or soft calfskin, sewed up like the sleeve of a coat or leg of a stocking, made twelve or fifteen inches long, about the size of the well hole, and open at both ends; this is slipped over the tube, and one end securely wrapped over knots on the tube to prevent slipping.
>
> Some six or eight inches of the bag are then filled with flax-

[12] A. Norman Tate, *Petroleum and Its Products*, 10–11.

seed either alone or mixed with gum tragocenth; the other end of the bag is then wrapped like the first, and the tube is ready for the well. When to their place—and they are put down any depth to hundreds of feet—the seed and gum soon swell from the water they absorb, till a close fit and a water-tight joint are made.[13]

13 Peckham, *op. cit.*

4

CONTEMPORARY ACCOUNTS

OF THE OIL REGION

THE excitement attendant to oil discovery did not stop at Oil Creek. It extended down the Allegheny River past Franklin and up to Tideoute. It was soon felt on French Creek and Two Mile Run. In 1864, discoveries along Cherry Run furnished the next excitement. This was followed by activity along Pithole Creek, the Bennehoff and Pioneer Runs, and the Woods and Stevenson farms along Oil Creek. Activity along Dennis Run and on Triumph Hill near Tideoute, near Shamburgh on Upper Cherry Run, and Pleasantville, during the next four years, contributed to the rapid expansion of stock companies and capital investment for exploratory purposes in the region.[1]

In October, 1860, a visitor to Oil Creek wrote: "So much oil is produced it is impossible to care for it, and thousands of barrels are running to waste in the creek. The surface of the river is covered with oil for miles below Franklin. Fears are entertained that the supply will soon be exhausted, if something is not done to prevent the waste."[2] At this time daily production was estimated to be 892 barrels; 450 barrels from wells along Oil Creek, and 442 barrels produced along the Allegheny River. A reporter from Franklin compiled the following statistics on the location of producing wells, their depth, and daily production.[3]

[1] J. H. Newton, editor, *History of Venango County, Pennsylvania*, 216–90.

[2] *The Derrick's Hand-Book of Petroleum: A Complete Chronological and Statistical Review of Petroleum Developments from 1859 to 1899*, I, 24.

[3] The *Venango Spectator* (Franklin, Pennsylvania), November 21, 1860.

WELLS ON OIL CREEK

Parker's Farm

Rockwell & Doan	30 barrels	220 feet
Parker Well	15	220
Barnsdall, No. 2	20	168
Barnsdall, No. 4	20	185
Tanner, Watson & Co.	35	147

Watson's Farm

Lumberman's Co.	24	158
E. L. Drake, No. 1	12	69
E. L. Drake, No. 2	30	300

Robison Farm

Adams & Co.	4	225
Plumer, Getty & Co.	25	150

Griffin Farm

Crossley & Co.	3	240

Poff Farm

Savage & Co.	4	220

Steele Farm

Empire Oil Co.	12	250
Smith & Co.	4	330

G. W. McClintock Farm

Kier, Watson & Co.	7	182

Rynd Farm

Kier, Watson & Co.	6	192

John McClintock Farm

Kellogg, Watson & Co.	10	185
Alden, Chas. & Co.	8	182

Buchanon Farm

Rouse, Mitchell & Brown	30	202
Halderman-Fountain Well	3	——

H. McClintock Farm

Persons, Dewey & Co.	10	180

Brewer & Watson, No. 1	7	180
Brewer & Watson, No. 2	8	222
Wilkins & Co.	10	192
Calkins & Bro.	6	100
Shirk & Co.	20	190
Purdue	20	188
McKinley & Abbott	4	——

Clapp Farm

C. Bell, No. 1	12	100
C. Bell, No. 2	10	425
Plumer, Eveleth & Bissell	15	155

Graff & Co. Farm

Graff & Co., No. 1	10	200
Graff & Co., No. 2	17	130

The reporter mentioned that operators near Tideoute had met with considerable success by drilling in and near the river bed of the Allegheny. He added: "It will be remembered that a preliminary injunction was prayed for in the Supreme Court against these borers. The River has anticipated the action of all Courts by getting high and sweeping away fourteen derricks." He listed the following producing properties:

ALLEGHENY RIVER WELLS

Lawry & Ford	40 barrels	76 feet
Vosburg & Co.	20	130
Hequembourg & Co.	15	120
Watters & Jackson	30	119
Economites	15	110
Ferris, Lyon & Co.	25	158
Bartlett & Mead	25	115
Tideoute Island Oil Co.	30	67

Goodman's Farm, Near Hemlock

Fryburg Co.	6	176

Anderson's Farm

Erie & Allegheny Co.	40	277
Catheart & Co.	2	230

McFates' Farm

Lamb & McCrory	12	182

Russell's Farm

Rodgers & Co.	5	124

Renauf's Farm

Arnold, Drum & Co.	40	400

Downey's Farm

Phillips & Co.	25	230

Hays' Farm

Niblo Seneca Oil Works	25	210

Wilson's Farm

Horner, Clawson & Co.	25	202
Raymond & Underhill (not fully tested)	12	375

Brandon's Farm

Peterson Well	10	270

Three Miles Below Franklin

Hoover & Stewart	20	——

A visitor to Oil City in 1861 wrote:

At Tideoute there are some 200 wells in progress, and all the way there here, 30 miles by raft, one is not out of sight of derricks and wells, hundreds and hundreds of them. Here I am more in the heart of the oily dominions than elsewhere. I find that New Bedford and Nantucket, heretofore oildom, has been unsuccessful for several years past, and is coming here, with its millions of money and its hordes of vessel officers, to harpoon the old mother of all whales (earth) and draw her blubber by the force of steam, which must eventually injure whaling oildom very much.

Drills going everywhere. . . . Everywhere in sight barrels, mountain high, from steamers just in from Pittsburgh, towing boats and barges loaded with barrels. The common topic of conversation everywhere is oil, rock, drilling, oil shows, depth, prices, prospects for the future, rents of land, best sites, chances for rapidly made fortunes, what has been, and what will be. My opinion is, that the landowners are grabbers; with the disposition to bless themselves and curse the world generally. They require one-half, and some as high as five-eighths of the oil for ground-rent; but time will settle all these things.[4]

A year later a visitor from Philadelphia reported:

Wonderful, most wonderful, marvellous, most marvellous, are the stories told of the oil region! It is, in fact, another California; the same greed for getting suddenly rich, the same spirit of speculation, the same rapid rise from poverty to affluence, and the same abandonment of other pursuits. As I came here (to Corry) from Erie a few days ago, I was introduced by Judge Johnson of Warren to a gentleman named V. N. Thompson whose receipts from his oil wells were one thousand dollars per day and who only a few years ago was a poor man and a bankrupt. Another instance was that of the firm of Nobles and Dellameter whose daily income was said to amount to $3,000; sixteen hundred barrels per day being filled from their well! I also learned that the Farrell well, at Titusville, produced when started more oil than eight hundred whaleships had collected in a year at New Bedford! The whalers of New Bedford and Newburyport have come hither in crowds, and find it much more agreeable to earn good wages, and sometimes to acquire large fortunes, than to be risking their lives amid the storms and dangers of the treacherous deep.[5]

[4] Morris, *op. cit.*, 42–43.
[5] *Ibid.*, 49–50.

Flush production brought transportation problems and falling prices:

The first flowing well ever struck was on the McElhenny, or Funk farm, and was known as the Funk well. It was struck in June, 1861 and commenced flowing to the astonishment of all the oil-borers in the neighborhood, at the rate of 250 barrels a day. Such a prodigious supply of grease upset all calculations but it was confidently predicted that the supply would soon stop. It was an "Oil Creek humbug" and those who had no direct interest in the prosperity of the well looked day after day to see the stream stop. But, like the old woman who sat down by the river side to let the water run itself out that she might cross dryshod, she waited in vain. The oil continued flowing, with little variations, for fifteen months, and then stopped, but not before Funk became a rich man.

But long before the Funk had given out, the wonder in regard to it was overshadowed by a new sensation. Down on the Tarr farm, the Phillips well burst forth with a stream of two thousand barrels daily. The Empire well, close to the Funk, suddenly burst forth with its three thousand barrels daily, a figure which subsequent flowing wells tried to equal.

The owners were bewildered. It was truly too much of a good thing. The true value of petroleum had not yet been discovered, and the market for it was limited. Foreigners would have nothing to do with the nasty, greasy, combustible thing. Our own people were divided in opinion. Some thought it a dangerous thing to be handled at arm's length, while others set it down as a humbug in some way or other, of which the community should keep as shy as possible. The supply was already in advance of the demand, and the addition of three thousand barrels a day was monstrous, and not to be endured. The price fell to twenty cents a barrel, then to fifteen, and then to ten. Coopers would sell barrels for cash only, and refused to take their pay in oil, or in drafts on oil shipments.

Finally it was impossible to obtain barrels on any terms,

for all the coopers in the surrounding country could not make barrels as fast as the Empire could fill them. The owners were in despair, and tried to choke off their confounded well, but it would not be choked off. They then built a dam around it, and covered the soil with grease, but the oil refused to be damned, and rushed into the stream, making Oil Creek literally worthy of its name. For nearly a year it flowed, and then dropped to a pumping well, yielding about a hundred barrels daily. Lately it stopped, but on the application of an air-pump it revived, and now runs about fifty barrels daily.

The Sherman well, which was the next great flowing well, was put down in the year 1862. It was sunk under great difficulties. J. W. Sherman, who was the original owner, commenced sinking it on the Foster Farm, next above the McElhenny, with very limited means. His wife furnished the money, and the well was sunk under great difficulties.

After a while it became necessary to procure an engine, and there was no money to make the purchase, and two men who were in possession of the desired article, were admitted to a share for the engine. Soon after, when but a few more feet were necessary to reach the supposed deposit of oil, the funds were exhausted. A sixteenth interest was offered for $100, and refused. Ultimately, it was sold for $60 and an old shot-gun. A horse became necessary during the work, and a share was bartered for the animal. At last, when all means that could be raised by borrowing or selling were about exhausted, oil was struck, and flowed at the rate of 1500 barrels a day. The flow continued at this rate for several months, when it declined to 700 barrels. For twenty-three months the well continued flowing, and then it stopped. For the first year, the proprietors made but little owing to the low price of oil and difficulty of getting it to market; but during the second year the market improved and an immense fortune was made. The well now pumps from thirty to forty barrels daily.[6]

The effect of oil upon the growth of Titusville was noted

[6] *Ibid.*, 39–40.

by a political reporter who visited there during a political campaign in 1863:

> Titusville, Crawford County, heretofore an inconsiderable village with a population three years ago of not more than four hundred inhabitants now contains nearly four thousand, and is realizing more wealth, by reason of the oil products alone, than California is sending here in gold. The habits of the people of Titusville and vicinity had become so settled, and they had so long lived an isolated life, that even after the great discovery in the oil region they could not conceive that a railroad could be built to tap that wonderful district, and to open it up to the commerce of the world; and they were accordingly startled from their propriety when, on the first of October, 1862, a locomotive dashed into their midst, dispelled prejudices, awakened their energies, and taught them that they were the citizens of a progressive world. . . . To show you how Titusville has grown, apart from the increase of population, I need only tell you that eighteen hundred two-horse teams arrive and depart every day from and to the oil wells.[7]

A correspondent from the *London Morning Post* visited the oil regions in 1863 and the following January wrote a detailed description from Oil City:

> As evening closed in, the office and public room of the Sheriff House gave abundant proof of the prevalence of the mud. Weary men entered in quick succession, all wearing long-legged boots, and all plastered and spattered with mud from head to foot. Sharp, keen-eyed men were they mostly, shrewd financiers and enterprising business men from New York and Philadelphia, with here and there a staid, cautious merchant from Boston, looking at every thing several times before making up his mind to invest, and then generally find-

[7] *Ibid.,* 50–51.

ing his more active and less cautious competitors from the other cities ahead of him. Adventurous speculators from the West, buying up lands with apparent recklessness, and then swearing at a handsome profit. Old Californians, familiar with the rush, excitement, and crowds of the early gold discoveries, but standing amazed at the greater rush and excitement of the oil diggings. People anxious to buy oil territory, and people with oil territory to sell, all crowded into the limited space occupied by the public room of the Sheriff House, and all muddy, dirty, and excited about oil. Every chair was soon occupied, and those unable to obtain seats leaned against the office-counter or the wall. The first question asked by the newcomer was invariably, "Can I get a bed to-night?" and the answer as invariably was, "Don't think you can. Will see what I can do for you by and by." With this all had to be content, and the next proceeding was to look out for a vacant chair, drop into it, and commence talking oil. No introductions were needed. Every one considered himself privileged to seek information from any person in the room, and the inquiries were taken as a matter of course, and courteously answered. Buying and selling went on without cessation. A gentleman from New York was describing to me a piece of property he had that day bought, on one of the tributaries to Oil Creek, for $10,000, when my right-hand neighbor, whose feet were planted next mine on the circular stove, and who appeared as if dozing, suddenly brightened up and inquired the exact locality of the property. Plans were produced, title-deeds examined, and in less than half an hour the property was resold for $14,000, the seller appearing very doubtful about the wisdom of his step. An incessant talking was going on all over the room, in which oil, oil, oil, was repeated with the monotonous iteration of the ticking of a clock. "A hundred thousand dollars—oil—struck a hundred barrel well yesterday—oil—flows two hundred and sixty barrels a day—oil— got nearly to the third sand-rock—oil—been offered three millions to sell out—oil." Such were the disjointed fragments of the general conversation that met my ear as I sat toasting my bootheels—American fashion—against the iron stove.

The large amount of the figures mentioned at first sounded like bombast; but a few inquiries as to the nature of the property bearing such high prices soon dissipated that impression, and familiarized me with the expanded ideas on money matters prevalent in the oil regions. When it is borne in mind that a well, producing a hundred barrels daily (and there are several wells on Oil Creek largely exceeding this product), yields a daily income of a thousand dollars, with no expense in the case of a flowing well, and but about ten dollars a day if a pumping well—the original expense of sinking the well, including the cost of engine and pumps, if needed, being not over six thousand dollars—it is evident that the purchase of even a part interest in such property will require a large sum. When, too, it is considered that there is room for several such wells on an acre of ground, it will be seen that the mere possibility of making such an oil-strike greatly enhances the value of land.

That night I slept in a bed. I was a privileged individual, and as, candle in hand, I followed the booted landlord up stairs to my resting-place, I was scowled on by more than one unfortunate who had passed two or three nights in arm-chairs or on the floor, and who considered themselves swindled out of their turn at the bed by the favoritism shown me. My couch was one of ten, each destined for two occupants, leaving but little unoccupied space in the room. There was but little trouble about ventilation, for the house had been built in a hurry, and was either not quite finished, or was in process of much-needed repair. The windows would not shut tightly, and several holes in the ceiling admitted air from the partially unroofed attic. An attempt to wheedle or bribe the landlord into giving me complete control of the bed miserably failed. The other man had the prior right, and I had been admitted to a half share as a mark of especial favor. My bed-mate came late, and persisted in giving me an account of the trials and difficulties he had experienced in sinking his well; but, said he, "there's a big show of oil, and by next week I expect she'll be worth $100,000 to me." There he stopped, and I soon fell asleep, lulled by the music of nineteen nasal organs, each playing with the bellows full of wind and all the stops out.

Next morning I was up betimes, and, booted to the hips, started on foot up Oil Creek. Derricks peered up behind the houses of Oil City, like dismounted steeples, and oil was pumping in the back-yards. Every foot of land on the creek is considered good boreable territory, and one reason alleged by the inhabitants for not improving the town is the fact that some day the houses will be torn down and the streets bored in search for oil. Since the formation of the town, three years ago, before which its site was a barren field, surrounded on all sides by a dense forest in which deer ran at will, down to the present time, when it numbers over 6,000 inhabitants, it is said but one funeral has taken place. There was, and is, no cemetery or burying-place, but the exigencies of a new country make Americans less particular than Europeans about the place of sepulture, and so the dead man was buried in a convenient lot. Unfortunately, that lot was sold in a few days as oil territory, and the body was removed to another place. The second place of burial was also sold as the site for an oil-well, and the body was at length shipped to Rochester, New York, to prevent its being bored through in the search for oil. Such was the tale narrated to me by "the oldest inhabitant," who claimed to have helped to ship the perturbed remains away, and its authenticity was vouched for by several others.

Oil Creek is a shallow stream, with a depth varying from a few inches in the dry season to about four feet after the spring and autumn rains. From Titusville, the present head of the actual oil-producing portion of the creek, to Oil City, is about twenty miles; and along the whole of this distance the ground is punched full of holes, and on most of it the derricks stand as thick as trees in a forest. Steep bluffs, ranging from 300 to 600 feet high, bound the narrow valley on either side, receding in some places to give room for less abrupt banks. Derricks throng the low marshy bottom-land, derricks congregate on the sloping banks, derricks even climb the precipitous face of the cliffs, establishing a foothold wherever a ledge of rock projects or a recess exists. The multitude of small wooden engine-houses, with their attendant derricks a few feet distant, look at a short distance like little old-fashioned churches, with detached campaniles, assembled in mass convention to discuss questions of theology or architecture. Here and there among the crowd were flowing wells, running from 100 to 1,000 barrels of oil a day, averaging $10 a barrel at the wells. Pumping-wells, forcing up from five to fifty barrels daily, were scattered thickly along the valley; the remainder of the derricks being over abandoned wells, or marking the situation of wells in progress. The air reeked with the scent of petroleum and gas, the mud under foot was greasy and slippery, the standing pools had the appearance of pure oil, and even the water of the creek was hidden beneath a mask of gorgeous hues formed by the waste oil floating on its surface.

Not a bright green thing was to be seen in the valley; every thing was black, sooty, and oily. The rocks were black and greasy. The trees that clothe the hill-sides were dingy with smoke, and the coarse grass and evergreen shrubs that struggled for existence, looked as if they had been drawn by a sweep through a sooty chimney. Mud was everywhere—deep, unlimited, universal; yellow mud up the newer territory of Cherry Run, dark mud on Oil Creek, dark green, repulsive mud in the vicinity of flowing wells. The prevalence of the fashion of wearing long-legged boots is explained, when every step sinks the leg nearly to the knee, and when the mud persistently endeavors to perform the office of a boot-jack.

Mud is a great leveller of distinctions. The wealthy capitalist of New York, the still more wealthy oil prince of the creek, and the laborer, working for the modest sum of five to ten dollars a day, can scarcely be distinguished from each other under the mask of mud worn by each. The millionaire sports

53

as muddy boots, as old garments, and is altogether as shockingly got up in the matter of dress as the man who runs the engine, or he who fills the oil-barrels. The barriers of social position are broken down in other matters besides dress. There is no assumption of superiority, no stiffness or reserve on account of wealth or station. On Oil Creek all meet on common ground. How can it be otherwise when Fortune, the fickle goddess, is so whimsical in the distribution of her good things in oil? The richest members of the oil aristocracy were three years since as poor as church mice. The men, who now possess millions, could not then, many of them, buy a second suit of clothes. Strange freaks of fortune occur every few days. An Irish laborer, employed in handling freight at the railroad depot in Titusville, a month or so since, invested $300 of his savings in a patch of four acres on the creek, near Titusville. An oil-well has been struck in the vicinity of his land, and to-day he was offered $5,000 for his purchase. He refused the offer, holding the property at $10,000, and he will get it.

One of the most successful, and at the same time most liberal, of the original operators in oil on the creek was Captain Funk, a man of limited means, but whose speculations in oil-wells built up for him a fortune of nearly two million dollars before his death, about a year since. Not long before his death a young man came to the creek in search of employment, and, when his money was entirely gone, applied to Captain Funk, who engaged him as clerk, at a moderate salary. Being well pleased with the manner in which his duties were performed, Captain Funk gave him an interest in a well he was then sinking. The well was a success, and a few days since the young man sold out his interest for $100,000. Still another young man, not yet twenty-one, was employed at wagoning oil on the creek. In part payment of his services, he took a sixth interest in a well then sinking. The well struck oil, and recently the young man sold out his property for $150,000, with a part of which money he purchased a large farm in Minnesota. These are but a few of the many instances of fortunes rapidly made that came under my notice. They are not exaggerations, or mere hearsay stories, but were told me by the parties most directly concerned.

A man named Taylor, who kept a lager-beer saloon in Cleveland, became affected by the oil fever, and started for the oil region with $700, his entire wealth, and a letter of introduction to an oil-dealer. He stayed two months waiting for some thing to turn up, and at last his opportunity arrived. An undivided thirty-second interest in a farm, on which a well was going down, was for sale at $1,500. He had $600 of his capital remaining, and the oil-dealer agreed to lend him the amount necessary to make up the purchase-money. A few days after the purchase was made the well reached oil. The value of the property suddenly increased, and Taylor sold out his thirty-second interest for $27,000. He is now operating largely in oil and oil lands, and has increased his capital enormously. Still another example of a similar character can be cited of a citizen of Cleveland, a Methodist minister, named Van Vleek. He was so exceedingly poor, that his family was in absolute want, and he had neither money nor credit to supply their needs. At this juncture Mr. Streater, the contractor for the construction of the Oil Creek railroad, employed him to negotiate for the right of way for the railroad. Van Vleek found an obstinate farmer, who refused to give the right of way, but offered to sell his farm at a moderate price. Finding him deaf to all other propositions, Mr. Streater told Van Vleek to buy the farm, and he would advance the money, which was done. Oil was afterwards found on the property, and Van Vleek sold out the farm, realizing a profit of more than $100,000.

But, in my journey up the creek, I found, connected with the history of many of the principal wells, instances of sudden elevation from poverty to wealth that would scarcely be credible, were it not for the tangible evidence before the eyes of every one. Standing before a flowing well, gushing out oil at the rate of $6,000 a day, without other expense to the proprietors than the original sinking of the hole, and the hire of two or three men to take care of the oil, it was easy to believe any story, however marvellous, of rapidly acquired riches.

The success of past operations in oil had stimulated adventure to such an extent, that a strong and general excitement reigned throughout the oil region, and, indeed, over the whole country. New wells were going down in all directions, the older derricks being jostled on all sides by the new-comers. The plan of seeking and boring for oil is of sufficient interest to warrant description.

It is unnecessary to enter into a full description of the manner in which the oil is produced in the bowels of the earth, or how it is distributed. My principal objection to doing so lies in the fact, that I am ignorant of the real facts of the case, and the several geologists, and other learned men who have made the matter their study, and to whom I applied for information, differed so widely in their explanations that I considered them not much better informed than myself. A plausible theory, and that most in favor with practical men, is that the present oil-beds were once salt marshes, covered with rank and salt vegetation. The subsidence of the earth's surface covered this vegetation with layers of sand, which, in process of time, hardened into sandstone, holding the vegetation prisoner. The hermetical imprisonment of the vegetation prevented its decomposition in the ordinary manner, and it was slowly distilled in the rock alembic. The component parts into which it was resolved—salt water, oil, and gas—gathered in the cracks and cavities of the rock, where it lay, awaiting for ages its release from bondage, by the operation of the miner's drill. From their difference of specific gravity, it is assumed, that the water lies at the bottom of the cavity, the oil next above, and the gas over all. If this was the invariable arrangement, and the cavity was of regular shape, the drill, in striking a cavity, would first liberate the gas, then the oil would have to be pumped out, and in the end the pump would draw nothing but water. The very many departures from this rule is accounted for on the hypothesis, that from the irregular shape of the cavities, the drill might first penetrate the middle, or oil-section, when the rush of gas to the hole would force up the oil, and thus cause a flowing well. After the gas had escaped, the flow would cease, and the remainder of the oil could only be extracted by the pump. The phenomenon of an exhausted well being revived, and yielding oil in paying quantities, after a short period of inaction, is accounted for by the theory, that in some instances several cavities are connected by minute channels through which the oil forces its way into the cavity first exhausted. Such is the most plausible of the many theories advanced; but even this is sometimes discredited by some of the new phenomena always making their appearance in oil-mining. An old whaling captain, living at Oil City—and, by the way, the number of whale-catchers and former dealers in whale-oil now engaged in the petroleum business is somewhat remarkable—explains the deposit of oil by the hypothesis, that a large shoal of whales were stranded in Western Pennsylvania, at the time of the subsidence of the flood, and that the oil-borers are now sinking holes in the blubber. His theory is not generally accepted, save among brother whalers.

With regard to the most favorable positions and conditions for sinking wells, there is a still wider difference of opinion. Arguments in favor of the flat bottom-land, the sloping banks, where there are any, the edge where the steep hill meets the valley, and even the hill-sides themselves, are each supported by instances of paying wells. In fact, there appears to be about as reliable a rule for selecting the location of a paying well as there is in deciding which is the fortunate number in a lottery. More than one well have been put down in obedience to dreams, the dreamer sometimes making a lucky hit, and at other times securing the blank of a dry hole. The plan, well known in English mining districts, of determining the proper spot by the witch-hazel has been introduced, and one or two lucky guesses having been made by the professors of the art, there is an increasing demand for their services. The oil-finder starts out in the morning with a forked twig, holding one end of the prongs in each hand, and having the pointer at the junction of the breast. In passing over the vein or basin of oil, it is the creed of the believers in the faith of the witch-hazel that the twig suddenly and irresistibly reverses its position, and points towards the earth. There the professor stops, and there the well must be sunk. As the same method has been adopted for years in discovering hidden springs of water, the borer not unfrequently obtains a copious yield of water instead of oil; but such accidents will happen in the best-regulated systems of science, and they only strengthen the faith of the true believers. Then there are oil-smellers, who profess to be able to discover the proper place for a well by smelling the earth.

The proper location being in some way determined, a derrick is erected, being a four-sided frame of timber substantially bolted together, making an inclosure about forty feet high, and about ten feet square at the base, tapering somewhat as it ascends. A grooved wheel or pulley hangs at the top, and a windlass and crank are at the base. A few feet from the derrick a small steam-engine is stationed and covered with a rough board shanty. A pitman-rod connects the crank of the engine with one end of a large wooden walking-beam, or engine bob, as it is called in some of the English mining districts, placed midway between the engine and derrick, the beam being pivoted on its centre, about twelve feet from the ground. This part of the machinery is of very rude construction. A rope, attached to the end of the beam nearest the derrick, passes over the pulley already mentioned, and terminates over the intended hole. A cast-iron pipe, from four and a half to five inches in diameter, is driven into the surface ground, length following length until the rock is reached. In the wells first sunk a pit was dug to the first rock, and a wooden tube put down. The earth having been removed from the interior of the pipe, the actual process of boring or drilling is commenced. Two huge links of iron, called jars, are at-

1. Samuel M. Kier (Drake Museum).

2. Colonel Edwin L. Drake, *inset* (Drake Museum).

3. Colonel Drake (in silk hat) in front of the discovery well in 1861 (Drake Muse

4. Replica of Drake well at Titusville, Pennsylvania (Cities Service Company; photograph by Fritz Henle).

5. Marker on spot where Drake well was drilled in 1859, *inset* (Cities Service Company; photograph by Fritz Henle).

6. Tools used on Drake well (Cities Service Company; photograph by Fritz Henle).

7. John Benninghoff Run at Oil Creek, Pennsylvania, in 1865 (Drake Museum).

8. Pithole wells at Balltown, Pennsylvania, in 1865 (Drake Museum).

9. Triumph Hill, Pennslyvania, in 1871 (Standard Oil Company, N.J.).

10. Spindletop, in 1929 (Standard Oil Company, N.J.).

State Capitol grounds, Oklahoma City (Standard Oil Company, N.J.).

12. Wells in Scurry County, West Texas, in 1950, *below* (Cities Service Company; photograph by Fritz Henle).

13. 1916 gusher at Glenn Pool (Standard Oil Company, N.J.).

14. "Christmas Tree" keeps flow of oil under control and thus avoids w
(Cities Service Company; photograph by Tony Lin

15. Wells drilled close together without a spacing plan (Standard Oil Company, N.J.).

16. Wells drilled according to forty-acre spacing plan (*The Oil and Gas Journ*

Spring-pole drilling in 1859 (from John J. McLaurin, *Sketches in Crude Oil*).

A modern drilling bit (Cities Service Company; photograph by Fritz Henle).

19. First oil "shipped" to Pittsburgh (from John J. McLaurin, *Sketches in Crude Oil*).

20. Bad roads near Oil Creek in 1863 (Drake Museum).

21. Pond freshet disaster in 1864 (Drake Museum).

22. Loading oil on barges about 1867 (Drake Museum).

tached to the end of the rope. A long and heavy iron pipe is fixed at the end of the lower link, and in the end of this is screwed the drill, or punch, a chisel-shaped piece of hardened steel, about three inches in diameter, and two to three feet long. When all is ready, the drill and its heavy attachments are lowered into the tube, and the engine set in motion. With every elevation of the derrick end of the walking-beam, the drill strikes the rock, the heavy links of the jars sliding into each other, and thus preventing a jerking strain on the rope. The rock, as it is pounded, mixes in a pulverized condition with the water constantly dropping into the hole, and assumes a pasty or slimy consistence. After a short time the drill is hoisted out, and the sand-pump dropped into the hole. This is a copper tube, about five feet long, and a little smaller than the drill, having a valve in the lower end, opening inwards. When the tube is dropped into the hole, the slimy fluid enters the tube through the valve, and is hoisted out. As the drill is chisel-shaped, the hole made by it would not be round unless some contrivance was resorted to in order to secure that end. This is accomplished in part by the borer, who sits on a seat about six or eight feet above the hole, holding a handle fixed to the rope, and giving the latter a half-twist at every blow. When the drilling is accomplished, another tool called a reamer is inserted, which makes the hole round and moderately smooth.

When the hole gets down to the point where the first reliable indications of oil are reached, the contents of the sand-pumps are carefully examined. The principal features of the geological formation of the Pennsylvania oil region are three strata of sandstone, with intervening strata of soapstone and shale. Indications of oil are found in the first and second sandstone, but the principal deposit is found in the third sandstone, at depths varying from three hundred to eight hundred feet. Should no oil be found in the third sandstone, the attempt is abandoned.

It will be seen that the process of drilling or punching for oil is extremely rude and tedious, sometimes occupying two or three months before reaching the average depth of five hundred feet. It is a matter of surprise that some ingenious Yankee had not long ago devised a quicker and better method of making a hole in the ground. A new plan has just been introduced, and I had the satisfaction of witnessing its operation at the experimental well, which has gone down about ten feet into the first rock. This is in reality, as well as in name, a drill, being a thin circular tube, set around the edge with six small stones of a species of diamond. A powerful engine with a proper machinery causes the drill to revolve with great velocity, cutting down into the rock and leaving a central core standing, which is jerked out with clamps let down for the purpose. The sections of the core brought up exhibited the formation of the rock in a distinct manner, giving a better idea of its character than had previously been obtainable. Should the experiment be successful, it will greatly facilitate the sinking of wells, as they can be sunk in a few days instead of weeks, and at no greater cost to the well-owners.

The well having been bored to the required depth, it is next tubed, an iron pipe with a valve at the lower end being run down the whole depth of the hole, the necessary length being obtained by screwing the joints together. As soon as tubed, and sometimes before, the oil and gas, should it be a flowing well, rush out with great force, and frequently with considerable noise. A pipe is connected with the upper part of the tube, and the oil conducted into an immense vat, holding from five hundred to one thousand two hundred barrels. The gas escapes into the air. If the oil should not flow, a pump-box, with sucker-rod of wood, is inserted in the tube and connected with the walking-beam of the engine. In many instances oil, gas, and water are all pumped up together, and are separated by a simple contrivance. The mingled fluids and gas are pumped into a small barrel. The oil and water fall into the bottom of the barrel, and run off by a pipe near the bottom into a large vat, where another separation takes place, the

60

greater specific gravity of the water causing it to sink to the bottom. The gas escapes by a small pipe at the top of the barrel, and is conducted into the furnace of the engine, where it burns with a fierce and steady flame, frequently dispensing with the use of other fuel.

It not unfrequently happens that a flowing-well degenerates into a pumping-well, and at length ceases to yield at all. In such cases the well has to be abandoned. A few months since a modification of the air-pump, under the name of an injector, was introduced, and has succeeded in restoring exhausted wells, and bringing them back to a paying yield. The principal feature of the contrivance is the introduction into the well of two small tubes, extending to the bottom of the hole, the smaller turning up at the bottom and entering the expanded end of the larger. Air is forced down the smaller tube with strong pressure from the engine, and is soon followed by a discharge of oil from the other tube, the volume of the discharge, in the experiments already made, gradually increasing. Proprietors of exhausted wells have thus found themselves unexpectedly restored to a handsome revenue.[8]

[8] *Ibid.*, 115–29.

5

TRANSPORTATION AND SPECULATION

A LITTLE MORE than a year after Drake's discovery well, the Oil Creek Railroad was chartered with authority to "build and construct a railroad from some point on the Philadelphia and Erie Railroad to Titusville, thence along Oil Creek to Oil City, and extend same to Franklin."[1] This road was commenced and completed early in 1862, from Corry to Titusville; the following year it was extended six miles down the creek to "Shaffer Farm." Meantime, before the road was extended to Oil City, transportation of crude oil from the wells along Oil Creek was dependent upon teamsters' wagons traversing almost impossible roads, or upon shallow-draft, flat-bottomed boats on the creek.

Along the courses of tributaries to the creek, lumber mills had built low-water dams for the storage of water used in manufacturing purposes. Oil producers effected a plan whereby the retardation dams could release a flood of water, locally referred to as a "pond freshet," when the creek was at low level, on which the flat-bottomed boats with their cargoes of oil could be floated along to its juncture with the Allegheny River.

A visitor to Oil City in May, 1862, wrote the following description of a pond freshet:

> In company with a nice party we paid a visit to Oil City on Saturday last to witness the last pond freshet of the season.

[1] Titusville *Morning Herald,* February 15, 1869.

To those who have never witnessed the operation of running oil on an artificial freshet, a brief description will be necessary.

The water during the week is collected in some twelve or fifteen dams, which are discharged of their contents when the boats are ready to run, and down they come, helter skelter, all conceivable kind of craft, with oil in bulk, oil in barrels, oil in all shapes. The crowd on the bridge at the mouth of the creek and the crowd on shore was immense. The sailors on the boats were the busiest beings we ever saw. Down comes the fleet, in good order, until one unfortunate strikes the pier of the bridge and swings athwart the current. That fellow loses his oil, which is in bulk, and blocks up the channel. The rear boats see the danger of collision but on they come—bump—crash—against the unlucky boat, which is soon reduced to chips.

From a cursory view of the spectators, and a "cussory" exhibition on the part of the boatmen, we come to the conclusion that almost every man on the bridge and on shore knew exactly how to run a boat. There was no lack of directions and orders by the shore fellows, and no scarcity of brimstone expletives from the men on the boats. Of the hundred crafts which started in the race, about ninety came through safely. Others were more or less damaged.

The oil spilled in the operation of running out of the creek floats into the eddies below the mouth, where it is gathered by men who devote themselves to that special business.

Oil City is really a business place. It has grown up with the oil trade and is an index to business on the Creek. On the east side of the creek several handsome residences are in progress. The town proper is very irregular in its plan, and everything that addresses itself to the eye or nose is full of grease. The plank pavements are saturated with oil. The sides of the houses are oily—the hitching posts are greasy—the men who congregate upon the boxes in front of the shops and saloons are oily, and leave their imprint where they sit. If any man wants to see what he can see nowhere else in the world, we recommend a trip to Oil City during a pond freshet.[2]

Pond freshets on Oil Creek were used extensively in 1862 and 1863.[3] In October, 1862, more than 100 boats took part, but only a few made safe passage. In December, ice at the mouth of the creek broke loose and pounded against 350 boats containing 60,000 barrels of oil; more than 150 boats and 30,000 barrels of oil were lost. In March, 1863, a pond freshet brought 15,000 barrels of crude down the creek. Early in May, over 18,000 barrels of oil made the passage, but some of "the boatmen started too early on the rise; a general crash and the loss of many boats and some 3,000 barrels of oil was the result." Later in the month a freshet brought down 12,000 barrels of oil. "Some empty boats broke loose ahead of the water, causing a crack-up of boats and loss of some 500 barrels of oil." In July, 145 boats safely rode the crest of a pond freshet, delivering 20,000 barrels of oil to the Allegheny River. On August 10, some 1,500 barrels of oil were lost when a lead boat hit a pier of the bridge across the creek at Oil City, and other boats were wrecked in the pile-up.

The cost of shipping oil by barge on Oil Creek at this time varied from two to five cents a barrel, according to the distance to the Allegheny River.[4] The toll was collected by agents passing along the creek, who collected the fees assessed according to the capacity of the barge, before the impounded waters were released. The October 29, 1863, issue of the Oil City *Register* contained a statement by A. S. Dobbs, superintendent of pond freshets, on the cost of dams:

> I here give the cost of dams and other necessary expenses of a pond fresh this year: Kingsland's dam, $200; Pierce's, $12; Stanton's, $20; Lytle's, $10; Child's, Benedict's and

[2] The *Venango Spectator,* May 21, 1862.
[3] *The Derrick's Hand-Book,* I, 24–33.
[4] Oil City *Derrick,* August 27, 1909.

Rouse's, $29; Tallant's, $5; Newton's, $29; Langworthy's, $7; Tryon's, $20; Hyde Creek, $29; for superintendent, horse hire, and the services of two men to cut dams, one on each side of the creek, $20; making a total of $383. All these dams cannot, at all times, be had, for their owners will not cut them for us. Mr. Benedict will not cut his at any price, hence we cannot get the Rouse dam, which is just above it. In dry seasons a "fresh" cannot be had every week, and if the superintendent spends a week or two extra between freshets, collecting tolls along the creek, this must be added to the expense, so that, in some cases a "fresh" actually costs $400. The same dam we now pay Mr. Kingsland $200 for, costs us in the early part of the season $55, and at no time during the summer of 1862 more than $100. Other owners of dams have raised their prices, but not in the same proportion. The rise in the price of oil, by encouraging team and rail shipments, has greatly diminished the amount of oil run out of Oil Creek, while the cost of a "pond fresh" is fully three times as great as last year.[5]

Earlier that year, Dobbs described pond freshets and Oil City as follows:

A pond freshet is a temporary rise of water in the creek for the purpose of running out boats, rafts, logs, etc. The water rises high enough to run out boats containing sometimes five hundred, and in some few cases, seven hundred barrels of oil. There are usually from one hundred and fifty to two hundred and fifty boats on each freshet. It lasts from one to two hours, and is caused by letting the water out from seven to seventeen dams on the principal branches of the Creek, so that the water will all meet together, making quite a flood upon which from seven thousand to thirty thousand barrels of oil are run in boats to the river.

. . . The shippers and boatmen having been notified of the

5 Quoted in *The Derrick's Hand-Book,* I, 33–34.

day upon which the freshet is to take place, begin to make preparations several days previous to it. Boats are overhauled, put in order, and then are towed, by means of horses, to the point on the creek from which they intend to start. The boats are then loaded and everything made ready for the flood which is to waft them to the much desired harbor at the mouth of the creek. About the time the freshet is expected the boatmen stand ready to loose their lines. A cool, pushing breeze is the first sign of it, and soon after come the swirling waters. Inexperienced boatmen generally cut their boats loose upon the first rush of water. As a matter of course, their boats run ahead of the water, and get aground upon the first ripple or shoal. The creek being very narrow and the force of the current generally swing these boats across it, a jam, and not infrequently a great loss of oil and boats ensues, just from the inconsiderate haste of a few. The experienced boatman waits at his harbor, until the water commences to recede, then cuts his line loose and trusts himself to the mercy of the swift current, and comes into port upon the highest part of the rise. The current of a first class pond freshet will run at the rate of six miles an hour. An ordinary one about four miles, and a small one two miles and a half. If the boatman meets with no obstacle he soon anchors his craft at our wharf.

There are several points of the creek where formidable obstacles are interposed to vex the navigator. Among these are the pier at McClintock Bridge, and a pier to support the machinery of a well in the middle of the creek immediately below; the Forge Dam, through which is only a narrow passage for boats; the pier of the bridge at this place, and the bar at the mouth of the creek itself. One boat getting across the creek at either of these places is apt to cause a "jam."

The boats are crushed against each other, and being generally built very light are easily broken, and if loaded with bulk oil, the contents are poured into the creek. If in barrels, the boat sinks, and barrels float off, and the owner rarely recovers them again.

66

Once landed at our wharves, the boat is either unloaded, or if the water is in good boating stage, goes, after brief preparation to Pittsburgh.

Our town is quite lively during the evening after a pond freshet. Shippers are busy paying off the boatmen, the citizens of the creek are laying in a stock of the necessaries of life, and all is bustle and business. You see men dripping with the oleaginous product. Our hotels are filled to repletion with these greasy men who are supplying the light for the world. Oil is the only topic of conversation, and the air is redolent with its sweet perfumery. The next morning boats are unloaded, and again towed up the creek for the next freshet, and comparative quiet again reigns in our city.[6]

Before the first decade of the oil industry drew to a close, however, pond freshets were only a memory. Railroad extensions into the oil regions and pipe lines from wells to shipping points reduced, too, the great number of teamsters employed in the transportation industry. John J. McLaurin has summarized conditions under which teamsters worked:

To haul oil from inland wells to shipping-points required thousands of horses. This service originated the wagon-train of the oil country, which at its best consisted of six thousand two-horse teams and wagons. No such transport-service was ever before seen outside of an army on a march. General M. H. Avery, a renowned cavalry-commander during the War, organized a regular army train at Pithole. Travellers in the oil regions seldom lost sight of these endless trains of wagons bearing their greasy freight to the nearest railroad or shipping point. Five to seven barrels—a barrel of oil weighed three hundred and sixty pounds—taxed the strength of the stoutest team. The mud was practically bottomless. Horses sank to their breasts and wagons far above their axles. Oil dripping

6 *Pennsylvania Petroleum*, 227–29. This is a reprint from the *Warren Mail* of January 24, 1863.

from innumerable barrels mixed with the dirt to keep the mass a perpetual paste, which destroyed the capillary glands and the hair of the animals. Many horses and mules had not a hair below their eyes. A long caravan of these hairless beasts gave a spectral aspect to the landscape. History records none other such roads. Many a horse fell into the batter and was left to smother. If a wagon broke the load was dumped into the mud-canal, or set on the bank to be taken by whoever thought it worth the trouble of stealing. Teamsters would pull down fences and drive through fields whenever possible, until the valley of Oil Creek was an unfathomable quagmire.[7]

Visitors to the oil regions met continuous lines of wagon trains hauling empty barrels and supplies to the well sites, and returning with their cargoes of crude oil.[8] In January, 1865, J. H. A. Bone walked from Shaffer's Farm to Oil City, twelve miles distant. On the way, he noted the following:

Here became visible the usual system of transportation adopted for oil and fuel, which is flat-boating on the creek. Four horses abreast are attached to a flat-boat, which they haul up stream, the horse taking the middle of the creek. The bed of the stream is even and covered with loose flat shale rock, the water being up to the horse's belly. An Oil Creek flat-boat generally holds from eighty to one hundred barrels of oil, on which the freight up is from seventy cents to one dollar, freight on coal being in proportion. As the boats sometime make two trips a day, the business is highly profitable, though anything but pleasant, especially to the horses.

As we passed down the creek the weather was intensely cold, and the ice was floating down in large masses, but the unhappy horses had to wade up with their heavy loads, their bodies partially clad in icy coats of mail, and their tails mere

[7] McLaurin, *op. cit.*, 262–63.

[8] Oil City *Register,* January 26, 1865; see also W. E. Youle, *Sixty-three Years in the Oil Fields,* 11.

bunches of icicles. If it is borne in mind that these horses had to be from three to four hours in this icy water, without relief or rest, and that even saddle-horses have to wade the stream several times in making the journey, the short lives and the wretched character of the live-stock in that region will not be wondered at.[9]

During this early period, prices paid for oil at the well fluctuated widely. Agents of the purchasing firms used every known device to ferret out information about new drilling discoveries; buyers could depress prices by reports of large strikes and heavy stocks. Generally, when shipment was active and stocks low, high prices prevailed; when shipment was dull, producers sold on a depressed margin or attempted to hold stocks in aboveground storage for an anticipated price increase.

In September, 1861, the Phillips well on the Tarr farm flowed 4,000 barrels a day; the Empire well was already producing at the rate of 2,400 barrels daily. The output from these two wells would have more than supplied the total domestic and foreign demand. Other wells added to the mounting supply along Oil Creek and the Allegheny River, while on Black Creek in the township of Enniskillen, between Lake Erie and Lake Ontario in Canada, thirty-five flowing wells were producing by early summer, 1862. It was estimated that not less than 5,000,000 barrels of oil floated off on the waters of Black Creek during the spring and summer of 1862. Rumors of Confederate raids drove the price of oil down to twenty-five cents a barrel at Parkersburg, West Virginia.[10]

Fluctuations in the price of oil are shown in the following table:[11]

[9] J. H. A. Bone, *Petroleum and Petroleum Wells,* 75–76.

[10] J. T. Henry, *The Early and Later History of Petroleum,* 130–35, 232–33.

[11] S. H. Stowell, "Petroleum," in *Mineral Resources of the United States,* 48 Cong., 1 sess., *House Misc. Doc. 75,* 203.

MONTHLY AVERAGE PRICE OF BARREL OF CRUDE OIL
AT WELL

	1860	1861	1862	1863	1864	1865
January	$19.25	$1.00	$ .10	$2.25	$ 4.00	$8.25
February	18.00	1.00	.15	2.50	4.37½	7.50
March	12.62½	1.00	.22½	2.62½	5.50	6.00
April	11.00	.62½	.50	2.87½	6.56	6.00
May	10.00	.50	.85	2.87½	6.87½	7.37½
June	9.50	.50	1.00	3.00	9.50	5.62½
July	8.62½	.50	1.25	3.25	12.12½	5.12½
August	7.50	.25	1.25	3.37½	10.12½	4.62½
September	6.62½	.20	1.25	3.50	8.87½	6.75
October	5.50	.10	1.75	3.75	7.75	8.12½
November	3.75	.10	2.00	3.85	10.00	7.25
December	2.75	.10	2.00	3.95	11.00	6.50

Many marginal wells along Oil Creek were abandoned during 1861–62 because of the depressed price of oil, and in many instances their abandonment caused waterflooding of near-by producing wells. This became of serious concern to producers who, through experience, lost valuable producing properties from waterflooding. More than 50 per cent of production from wells was obtained during the first year, and producers, by the mid-sixties, became interested in doing everything possible to prolong the life of wells. For example, on May 27, 1865, producers and landowners of the Tarr farm met at Tarrville and decided every well should be cased down to the second sandstone, and seedbagged at that point, "it being, by all present, thought much better to seedbag at the second sandstone than at the first or third."[12] Expenses were borne at a pro rata amount.

Less than a year later, a visitor to the region reported:

During 1862 the average daily production on Oil Creek was 8,000 barrels but in 1863–64 scarcely exceeded 3,000 barrels

[12] Oil City *Register,* June 8, 1865.

daily. The reason for the decrease has since been satisfactorily explained. The wells were sunk in such close proximity, that they interfered by letting down the water or "drowning out" each other. A well left without being tubed, let down the water, frequently interfering with, or stopping a dozen of the adjoining wells.

Many, in fact, a majority of the wells bored that have proved unproductive have been abandoned, the tubing removed, and no provision made for keeping the water from going to the depths below, where the oil deposits are. Besides, the gas, which is the motive power so to speak, that forces the oil from its hidden depth in the earth to the surface, is thus allowed to escape through these abandoned wells and is lost. From an observation of several years we hazard the assertion that the causes alluded to above have produced the decrease in production in various localities, and is not occasioned by the exhaustion of the supply.

In proof of this assertion we cite the case of the Tarr and Blood farms, upon Oil Creek. The Woodford and Philadelphia wells on the Tarr farm were affected whenever the tubing was withdrawn from either. A majority of the producing wells on these farms were literally deluged with a flood of water from these wells. In 1865 the operators concluded to case all the wells with iron casing to the second sandrock and apply a seedbag at that point. The land and the working interests both combined in the movement. The water was then pumped out of the wells. The result has been wonderful. At the time the movement was inaugurated, the daily production of both farms hardly exceeded 150 barrels a day. The production during the past season and up to the present time has averaged over 650 barrels.

Petroleum, which is doubtless the accumulation of ages, is deposited in reservoirs of vast size. The lighter qualities of the oil, which constitutes the gas, invariably follows the perforation of the drill, and forces through the orifice thus made like the steam from a boiler. If properly used and economized the

71

period of its duration to effect the operations for which nature doubtlessly intended it, forcing the oil from its depths to the earth's surface, could not be well estimated. This gas is allowed to escape through the abandoned wells, and the motive power is thus lost. Without the assistance of the gas, the quantity of oil obtained by pumping is hardly appreciable. In fact the apparatus now used is about as adequate to pump oil from a depth of 500 to 750 feet as a hand pump would be to pump the Atlantic Ocean dry. There is no effort made to economize this motive power. Gas wells, as they are called, have been struck in several localities which were allowed to stand open until they exhausted themselves. The consequence of this procedure, we have noticed, is invariably a decrease in production of a neighboring locality. That oil exists in every locality where there has been any production, in practically inexhaustible quantities, is, to our view, certain. Without the use of the gas as a motive power, there has as yet, been no machinery devised that possesses sufficient power to bring it out of the ground in any great quantity.

We have briefly given our views of the cause of the decrease of petroleum production in localities. We are satisfied that time will prove its correctness. These views have not been hastily made up. We are confident that no appreciable impression has as yet been made upon the petroleum supply. If the same measures that have been adopted on the Tarr and Blood farms, together with such improvements as their natural use would suggest, be made general, the petroleum supply could not only be made of indefinite duration but it could also be safely regulated to suit the wants of the producers and the trade generally. For the present extravagance and waste that prevails the producers will have to pay a heavy penalty. We urge the necessity of the people of every producing locality to apply a speedy remedy to prevent water flooding and the gas from escaping as they thus lose a previous boon which kind nature has provided for their benefit.[13]

[13] Titusville *Morning Herald,* April 26, 1866.

Despite the unusual risks involved in investing in prospective oil properties, and contrary to the knowledge of practical oil men that more money was expended in drilling unproductive wells than those that produced, stock companies flourished during the first decade of the petroleum industry. It became a common practice for acreage to be obtained near producing properties, in order that the prospectus of the company could cite a well so many rods distant, to attract subscribers. In many cases, the prospective location attracted subscribers merely by being listed in the Oil Creek vicinity. By 1866, at least 460 companies were listed; however, the stock of no more than 15 was of any value; "of the remaining 445 companies we can say nothing promising, for they are so utterly worthless that their stock can not be disposed of at any price on the stock board."[14]

The speculative frenzy that attracted dreams of easily won wealth from the oil regions has been described by David Lowry:

> Heterogenous masses of humanity were wedged in the railway coaches that carried travelers from Franklin to Oil City. They were packed three in a seat, squatting on the seat rails and crowding the aisles in those days when the "Gripsack Brigade" (the convenient title comprehending all newcomers on Oil Creek), led by dazzling dreams of wealth, swept like a tide over Venango County.
>
> It required some time to become accustomed to the jargon of the Oil Region. Eighths, 12th and 16th royalties; quarter, 8th, 16th, 32nd and 64th interest in wells; options and bonuses —all, however, were terms pregnant with meaning. Sand-pumps, bull-wheels, bits, rimmers, jars and sucker-rods, soap-veins, nigger-heads, stuck tools—these and other terms made up a jargon employed with a liberality that was very discouraging to new-comers.
>
> The prospectus writer's art was concentrated upon oil. A forest could scarcely supply pegs sufficient to stake out the

14 *Ibid.*, May 2, 1866.

leases made by companies in Philadelphia. In Pittsburgh the
oil companies out-numbered the lamp posts. Shares in wells
were divided and sub-divided until housemaids and cooks,
barbers and porters boasted of 32nd, 64th, and 128th interest
in wells that never accomplished more than to disturb the
wood-peckers for a brief season. Eight and quarter interests
made men snug fortunes. Interests in wells considered separate and apart from the remainder of a lease were sold as
readily as interests in leases containing 5, 10, or 50 acres. A
10-acre lease in a favorable location was good for almost any
amount in the latter part of 1863, and throughout 1864 and
1865. Tens and hundreds of thousands of dollars were paid
for leases that could not have returned the purchasers 10
cents on the dollar if every acre had yielded 1,000 barrels for
a year. With oil dropping in price in 1866 and dry holes out-
numbering productive wells, the money sunk in leases, and in
lands purchased outright, speedily assumed proportions much
greater than the vast profits made by the fortunate few.[15]

Fabulous oil strikes in 1864, however, prompted greater
speculation. A flowing well was struck on July 18, up Cherry
Run on the Rynd farm, on a lease consisting of one acre of land;
three other producing wells were immediately developed on the
acre-tract and a one-fourth interest in the lease sold for $280,-
000.[16] The story of the Egbert and Hyde farm—a 38-acre ob-
long-shaped flat, across Oil Creek from the McClintock farm,
purchased in 1859 for the express purpose of drilling for oil—
was common knowledge. Not developed until 1864, it was estimated the thirty-four producing wells returned $6,000,000 to
the owners in twelve months' time. Flowing wells on the tract
included the Maple Shade, with 1,200 barrels daily production,
the Jersey, with 400 barrels, Keystone No. 1 and No. 2, each
producing 300 barrels, and the Coquette, 600 barrels. Egbert

15 "Oil Creek in the Sixties," in the Titusville *Herald,* August 22, 1934.
16 J. T. Henry, *op. cit.,* 231.

sold a one-twelfth interest in the latter well for $200,000. The price of oil at the well averaged more than $8.00 a barrel during this period of flush production. The product of the Jersey well, struck in the early part of the spring of 1864, sold as high as $14.00 a barrel, and made the highest average of any of the enormously profitable wells on the Egbert and Hyde farm.

Fowler described interest shown by the public when shares were offered for sale in the Maple Shade well.

> When the truth was ascertained concerning the yield of the largest well on the Egbert & Hyde farm, a company was organized in Philadelphia, with a capital of $500,000 and 50,000 shares were placed on the market; the par value was $10 a share. On the day the shares were disposed of, a line of applicants was formed at the office similar to those that besiege the box-office of an opera house or theatre when extraordinary attractions excite the public. Those within hearing of the office were seen holding their hands up, flourishing rolls of bank notes while they shouted their names and desired to be remembered in the sales.[17]

Within ten days, shares of the stock brought $20.00, and subsequently, $44.00.

A prospectus of a joint-stock company, chartered by one of the eastern states, could always incite public interest in investment if proven production was near-by the company's lease. Typical was the announcement of the People's Investment Company, which listed six wells on an adjoining property, then stated: "There is room for 30 or more wells on our 7-acre lease." The company had two producing wells on a two-acre tract and announced: "On these two acres there is room for seven or more wells."[18]

[17] "Oil Creek in the Sixties," *loc. cit.*
[18] "Prospectus of the People's Petroleum Company," 6–7. A copy is on file in the Drake Museum.

Annual reports to stockholders, too, during this period, were optimistically worded. One quoted a letter from the superintendent, who wrote: "I sold 1,000 barrels of oil this day at $5.25 per barrel; this will make $3.90 for the oil, after paying the dollar tax and the cartage of the oil to the creek."[19]

A few of the companies proved to be highly profitable organizations. Best known along Oil Creek was the Columbia Oil Company, which purchased the Story farm in 1862 for $128,000. It was organized with a capital of $200,000, divided into 10,000 shares of the par value of $20.00 each. During the first year, shares sold in the market for from $2.00 to $10.00, but with profitable producing wells, shares had risen to $125 each by March, 1864. An article in the Pittsburgh *Evening Chronicle* stated: "The person who paid one year and a half ago the original value of $20 for one hundred shares, and has held his stock, has received $12,000 dividends up to December, and from the profits on the increase in capital made in June last, obtained an accession to his stock of four hundred shares; which shares, although of fifty dollars par, are now worth, with his original 100 shares, $42,500, making a clear profit of $54,500 in eighteen months."[20]

On such successes was the appetite of a public whetted with war-inflated currency for investment. The craze for oil stocks and the tendency of many persons to be duped or gulled into investment brought the following comment from a newspaper reporter in the area:

> One of the most singular features of the times is the extraordinary demand that has sprung up for oil stocks. If this demand were confined to the stocks of producing companies we could understand it, but applying as well to the stock of

[19] "Second Annual Report to the Stockholders of the Central Petroleum Company, December 15, 1865," 4. The copy examined is in the Drake Museum.

[20] Pittsburgh *Evening Chronicle*, December 28, 1864, quoted in Document No. 128, *Pennsylvania Petroleum*, 252.

organizations which have not yet commenced boring even, as to the dividend paying shares, we are at a loss to comprehend it. Indeed such is the rage for stocks of this character now here that sales are common in companies where the stock certificates have not yet been issued or the transfer books opened, and instances are not uncommon where stock has been sold at an advance of two hundred per cent even before the company was organized![21]

By 1866, it was estimated more than $40,000,000 had been invested in joint-stock companies organized for oil exploration, most of which were organizational schemes with worthless stock, some of which were reputable and profitable, with shares regularly traded at stock-exchange centers. A writer, at the time, commented:

In many cases we know of parties who have sunk from $50,000 to $100,000 without the slightest hope of seeing one red cent again. A very heavy proportion of the loss has fallen upon parties with straitened financial means, and ruin has in many cases followed the attempt to make a fortune in a month by dabbling in bogus oil stocks. King Solomon says: "He that maketh haste to be rich shall not be innocent," and many unfortunate speculators have groaningly acknowledged the fact when they have seen their scheme fail, and when they have been arraigned either in court or at the bar of public opinion as common swindlers. The inside history of many "operations" on the "ground floor," etc., would be astonishing. Men of high standing in church and state would be proved to have been liars, hypocrites, reckless gamesters, and utterly devoid of all humanity or honesty. Other people, supposed to be shrewd business men, would be shown to have been the silliest dupes imaginable.[22]

[21] Pittsburgh *Evening Chronicle,* August 26, 1864, quoted in Document No. 125, *Pennsylvania Petroleum,* 245.
[22] Titusville *Morning Herald,* May 2, 1866.

Although speculation in oil stocks was to continue, the public became more wary of joint-stock promotional schemes for oil production after the national attention drawn to many failures in the Pithole area. Meantime, producers in the oil region were undergoing a revolutionary change in the transportation of crude oil, the use of pipe lines, and the adoption by railroads of cars to transport oil in bulk.

The introduction of pipe lines for the transportation of crude oil created more hostility in the oil regions than any other factor. Suddenly, thousands of teamsters engaged in hauling barrels of crude oil were thrown out of employment. Some resorted to sporadic acts of violence to obstruct the operation of the lines that interfered with their traffic.

A successful attempt to transport crude oil through cast-iron pipe was made in 1862. J. M. Barrows conveyed oil from wells in the Tarr farm one thousand feet to his refinery. That same year a bill was introduced into the Pennsylvania legislature to incorporate the Oil Creek Transportation Company, an organization "to carry oil in pipes or tubes from any point on Oil Creek to the mouth of that creek, or to any point on the Philadelphia and Erie Railroad."[23] Despite opposition by teamsters, the bill was enacted into law. Two years later the legislature granted a charter to the Western Transportation Company, which laid a line from the then famous Noble well to Miller Farm station on the Oil Creek Railroad, a distance of three miles. The previous year a pipe line two and one-half miles long connected the Empire well with the railroad terminus at Shaffer Farm. Both attempts to transport crude oil by this means proved unprofitable because of leaking joints, ruptured pipes, and vandalism from teamsters.

J. L. Hutchings came to the oil country in 1862, with a rotary pump he had patented. He proved that crude oil could be forced through pipes made leakproof at the joints and sufficiently strong

[23] Act of February 26, 1862, Pennsylvania Statutes.

to withstand the tremendous pressure and vibration from the flow. With the development of a better grade of pipe, screw joints, collar and sleeve at connections, and the use of the rotary pump to force the flow, the mechanics of carrying oil by pipe was established.

After the Pithole field was developed, Samuel Van Syckel, in August, 1865, completed a two-inch line from Pithole to Miller Farm, the railway terminus, four miles away. Two pump stations helped force some eight hundred barrels of crude oil daily to the shipping point.

This success prompted the formation of a company to transport oil by pipe from Pithole to Oleopolis, a shipping point on the Allegheny River. The laying of the pipe was commenced in November and the first oil was run through its seven-mile length in December. This was a gravity-flow line; the continuous downgrade from a fall of 360 feet permitted the pipe to drain itself. The line was six inches in diameter, with a capacity of 7,000 barrels.[24]

At the same time, Henry Harley and associates extended pipe lines, using two-inch tubing of extra weight, from wells along Benninghoff Run to storage and shipping points at Shaffer's Farm, about two and one-half miles away. Teamsters cut the pipes, burned the tanks, and retarded the work seriously. The daily capacity of these lines was 2,000 barrels. They were placed in operation on March 10. Six weeks later it was estimated they were carrying 1,500 barrels daily, yielding a revenue to the company of $750 per day.[25]

Teamsters could foresee their unemployment and resorted to violence. The issue of the Titusville *Morning Herald* for Thursday, April 19, reported:

> About two o'clock yesterday morning a fire broke out at Shaffer, in the tanks of Henry Harley and Company, pro-

[24] "Crocus" (Charles C. Leonard), *History of Pithole*, 37.
[25] Titusville *Morning Herald*, April 21, 1866.

prietors of the Bennehoff and Shaffer Pipe Company. The fire
started on the upper tank, which contained about 250 barrels
of oil. By the exertions of the employees it was confined to the
first two tanks. The oil, however, flowed down to the shipping
platform, entirely consuming it. Four tank cars, belonging to
Deane and Company, tank-line, together with 450 barrels of
oil on the platform and about 300 barrels in the tank cars were
consumed. The fire is thought to be the work of an in-
cendiary.[26]

Two days later the writer referred to the fire and reported
another:

On Friday morning about 2½ o'clock the operation was
repeated, under the auspices of a formidable mob, numbering
between seventy-five and one hundred men, armed with revolvers. They rushed down from the timber and underbrush
at the hillside, and before reaching the tanks halted, when the
ringleader notified the watchmen that they intended to destroy the tanks and called upon them to disperse; or they
would be shot down. The watchmen hesitated, but made no
reply. The mob then fired upon them, but without effect. The
watchmen retreated and discharged their own revolvers at the
mob, and it is supposed with effect. One man was heard to ex-
claim, "My God! I am shot."

Another of the mob was seen to throw a fireball into one of
the tanks, and in an instant the flames burst forth, communi-
cating of course to the adjacent tanks. There were five in all,
but fortunately the oil had been drawn down the previous
day, and there were only 800 barrels of oil in store.

. . . This process of moving oil has entirely superseded the
old method of hauling oil in barrels from Bennehoff to Shaffer,
and of course the teamster's occupation is gone. There have
been murmurings both "loud and deep" on the part of the
aggrieved, ever since the new line went into operation on the

26 *Ibid.*, April 19, 1866.

10th of March. The company were apprised of their danger, and established a patrol on the route of the pipes, and a relief guard at the receiving tanks; but they were unprepared for so formidable an onslaught as that of Friday morning. It is asserted the raiders are fully organized and disciplined, and have threatened to destroy the property of all in the region.

The tanks at Miller Farm, belonging to Mr. Van Syckel, and the tanks of Sherman and Pickett, at Titusville, it is alleged, are both marked for destruction. What course the owners of the property will adopt for its protection it is not yet determined.

Owners of pipe lines and storage facilities quickly took action to protect their property. In addition to the employment of watchmen at storage sites and along the lines, hired detectives joined the malcontents. Ringleaders were arrested and mob rule overthrown. By the end of the first decade of the oil industry, wells were connected by pipe directly with storage tanks at the pumping station, and whenever new fields were opened, pipe lines were extended to take care of the production.

Storage facilities and railroad cars to transport crude oil likewise improved in quality and capacity during this period. At the beginning of the period, storage other than dug pits consisted of wooden tanks constructed to hold a 200- and later a 1200-barrel capacity. By the summer of 1866, however, wooden tanks were rapidly being replaced by those of larger capacity built of iron. A 10,000-barrel tank was built at Petroleum Center in June; several were under construction at Oil City, the largest with a capacity of 14,000 barrels.

It was estimated that an iron tank of 12,000-barrel capacity was not so expensive as twelve wooden ones of 1,000-barrel capacity, and served as added protection against loss by fire. The tanks were built of boiler iron, riveted, and "secured with such perfection as to render them perfectly fire proof," according

81

to an article that appeared in the Titusville *Morning Herald,* on April 25.[27] The writer added:

> It is intended in this manner to insure the millions of dollars of property which are now annually destroyed by the fires and floods which so frequently devastate the oil regions. The present rates of insurance are so high as to be practically prohibitory, while with the use of these iron tanks, the proprietors can perform at a reasonable charge the double business of storage and insurance.
>
> The necessity for safe storage of petroleum in this region has long been apparent and the importance of an enterprise that will preserve this valuable commodity from the elements of fire and flood can scarcely be overrated. The large accumulation of oil from the excess of production must be held by someone. If it is held at the wells in wooden tanks by the producers, the danger and liability to be swept off at any moment by fire is very great; and if it is held in the eastern markets in barrels, the amount of money locked up in barrels and freight is enormous, to say nothing of the constant depreciation in value by evaporation. With the ample storage facilities proposed, operators in the eastern market could then buy oil here and hold it until they wished to sell it or fulfill their orders, or the producers could hold it without risk of losing it by fire, or excessive loss by evaporation, which cannot be avoided in wooden tanks.

Although the tanks were not fireproof—they were more fireproof and practicable than wooden tanks—they came into common use in the oil industry. By July, when daily oil production was estimated to be 11,000 barrels, the total capacity of iron tanks did not exceed the amount of oil recovered in a week; by the end of the year, storage capacity in iron tanks was estimated to exceed a half-million barrels. One company alone was build-

[27] *Ibid.,* April 25, 1866. Issues of April 16–17, June 2, and July 13, 1866, include articles on the iron-storage tank industry.

ing a 10,000-barrel tank every twenty days. By the end of the first decade of the oil industry, tanks with storage capacity of 20,000 to 30,000 barrels were not uncommon.

The first oil shipped by rail was contained in wooden casks or barrels. Leakage caused much loss. A flatcar could be loaded with sixty barrels, set end on end. During the summer of 1865, Amos Densmore of the shipping firm of Densmore, Watson and Company, conceived the idea of shipping oil in bulk.[28] In September, two cars of the Atlantic and Great Western Railroad were each equipped with two wooden tanks to hold the oil. Each car carried 84 to 90 barrels of bulk oil at the rate normally charged for 60 barrels. This first shipment, from Miller Farm on Oil Creek to New York, proved successful; other cars were equipped for the bulk shipment of oil, and by April of the following year it was estimated there were four hundred tank cars in use, capable of carrying 32,000 barrels of oil which not only saved capital formerly tied up in barrels but saved on carrying charges. The freight rate for an 80-barrel car was the same as that formerly charged for transporting 60 barrels.[29]

In 1869, one of the principal shippers of crude oil, the Empire Transportation Company, first put into use iron-boiler tank cars of eighty-barrel capacity. Knowledge of loss of contents due to expansion in wooden tank cars led to the adoption of a dome on each car to compensate for the expansion factor. Experiments proved this should be a ratio of one gallon per barrel of tank capacity; therefore, a dome was adopted of eighty-gallon capacity.[30]

[28] "The First Oil Tank Car," Document No. 141, in *Pennsylvania Petroleum*, 382.

[29] Titusville *Morning Herald*, April 12, 1866.

[30] "Empire Transportation Company and Oil Tank Cars," Document No. 151, in *Pennsylvania Petroleum*, 306–308.

6

THE "RULE OF CAPTURE"

BY THE CLOSE of the first decade of the oil industry, producer organizations had successfully defeated the effort of the federal government to keep in effect an excise tax on each barrel of oil produced. They were not so successful in the attempt to overthrow the monopolistic franchise on the use of torpedoes to stimulate production.

The federal Congress, in seeking to increase revenue to meet the costs of the Civil War, enacted an excise tax of one dollar per barrel of crude oil produced.[1] The tax immediately aroused strong protests from the oil-producing regions. In an editorial which explained the crippling effects the excise tax had imposed on the producers, the Oil City *Register* summarized the importance of the oil industry:

> It adds, moreover, to the national wealth, gives employment to railroad, mechanical, and shipping interests, and takes the place of gold in the purchase of European goods and the settlement of foreign claims. As an illuminator and lubricator it is but little more than half the price of sperm and whale oils, and its use is daily being extended in every department of mechanics. It is used largely in the manufacture of soaps and all the finer toilet articles, while some of the most beautiful shades of color ever known are being, by chemical

[1] "Internal Revenue Act of March 3, 1865," United States Statutes at Large, XIII, 484.

process, extracted from it. And besides all this, the residium is now being brought into use as a steam generator, to navigate the ocean, with most flattering promise of success. The cities of Moscow and St. Petersburg are lighted with petroleum exported from this country and it is destined, beyond question, to enter very largely into the manufacture of gas, thereby materially reducing the cost of that very necessary article.[2]

During the summer and fall of 1865, congressmen and senators visited the oil region, and no opportunity was lost at meetings in Pithole, Titusville, and Oil City to impress upon the legislators the onerous burden of the tax which adversely affected marginal-producing wells. One reported the "country was covered with abandoned derricks, monuments to departed greatness."

A meeting of oil producers was held at Titusville on Friday evening, April 20, 1866, to protest against the continuation of the tax. A resolution was prepared and forwarded to the state assembly at Harrisburg, to enlist the aid of the legislative body in memorializing Congress not to re-enact the law. A similar resolution was prepared for C. V. Culver, their representative in Congress. George M. Mowbray, a principal spokesman in favor of repeal of the tax at the mass meeting, was chosen to represent the producers at the nation's capital. Because of the organized opposition, the House of Representatives, on May 7, voted to repeal the tax; two days later similar action was taken in the Senate.[3]

The fight for unlimited manufacture of torpedoes and their use in exploding underground strata pierced by the drill was long drawn out, expensive, and a failure. As knowledge increased in regard to oil-bearing sands, so grew interest in increasing production from wells. Colonel E. A. L. Roberts, then an officer of

[2] January 26, 1865.
[3] Titusville *Morning Herald*, April 21, 23, May 3, 8, 10, 1866.

volunteers serving as federal troops in the Civil War, has been credited with the idea of exploding torpedoes in wells. Although rival claimants arose and proof was presented that the system had been experimented with before his appearance in the oil region, Roberts had the foresight to apply for a patent in November, 1864, and constructed six torpedoes according to methods he had projected.

Roberts came to Titusville the following January and persuaded the owner of the Ladies Well, on Watson Flats below the city, to agree to the experiment. The explosion in the well of two torpedoes at the depth of 465 feet successfully proved that fracturing the oil sand could result in greater production. The following year he exploded torpedoes in a dry hole near proven territory. This experiment took place at the Woodin well, on the Blood Farm near Oil Creek. The explosion resulted in a twenty-barrel well. This established the practice of using torpedoes.

Colonel Roberts and his brother, Dr. Walter B. Roberts, organized the Roberts Torpedo Company and during the next fifteen years engaged in an interminable number of lawsuits for infringement of their patent, which was reissued in 1873, and against producers who used explosives without paying a fee to the company.

On Friday, February 12, 1869, leading producers met at Titusville and formed the Oil Producers Association. Jonathan Watson of Titusville was elected president and Andrew Cone of Oil City, secretary. A board of managers consisting of thirty-five members from producing districts was chosen, and membership fees and annual dues provided for. The major action item of the meeting was set forth in an unanimously adopted resolution:

WHEREAS, The free use of torpedoes for the purpose of increasing the production of oil wells is now hindered and restricted in consequence of the claims by various parties of the exclusive right to use the same, or to make and explode them by the most convenient methods: and WHEREAS, The exclusive right to use casing for wells, and various improvements therein is in like manner claimed; and believing that such claims are not well founded, or if well founded that the interests of the public would be greatly promoted by procuring, by the most equitable and proper means practicable, a surrender of such claims; and WHEREAS, It is the duty of the oil producers to protect their interests against the impositions of unjust monopolies, therefore, Be it unanimously *resolved,* That the cooperation of the oil producers be invited in testing the validity of the claims above set forth.[4]

Producers subscribed $50,000 to break the Roberts patent. Their efforts were fruitless. Colonel Roberts, shortly before his death in 1881, stated he had spent $250,000 in torpedo litigation; he was responsible for more lawsuits than any other man in the United States.[5]

An early court opinion included a description of the process first used by Roberts:

The patent consists in sinking to the bottom of the well, or to that portion of it which passes through the oil-bearing rocks, a water-tight flask, containing gunpowder or other powerful explosive material, the flask being a little less in diameter than the diameter of the bore to enable it to slide down easily. This torpedo or flask is so constructed that its contents may be ignited either by caps with a weight falling down on them or by fulminating powder placed so it can be exploded by a movable wire or by electricity, or by any of the known means for exploding shells, torpedoes, or cartridges under water. When the flask is sunk to the desired position, the well is filled with water, if not already filled, thus making

4 *Ibid.,* February 13, 1869.
5 McLaurin, *op. cit.,* 336.

a water tamping and confining the effects of the explosion to the rocks in the immediate vicinity of the flask and leaving other parts of the rock surrounding the well not materially affected. The contents of the flask are then exploded by the means above mentioned, and, as the evidence shows, with the result in most cases of increasing the flow of oil very largely. . . . The explosion breaks obstructions and permits oil to reach the well.[6]

The Roberts Company soon abandoned the water-tamping method of explosion and developed more powerful explosives through the use of nitroglycerin. Two hundred dollars was received by the company for a medium shot. Because of the price and resentment against monopolistic control, there were many evasions or circumventions to escape paying the fee. Also, many parties in the oil regions manufactured nitroglycerin and exploded it in the wells by stealth, customarily delivering and setting off the charge at night. Those engaged in these hazardous undertakings were called "moonlighters." Another and safer method of avoiding full royalty payment was also followed. The owner of a well to be torpedoed would purchase two-thirds or three-fourths of the amount of nitroglycerin required from outside parties, viz., forty quarts of a sixty-quart charge, and then engage the Roberts Company to put in the other twenty quarts and supervise its use. These were called "setters"; they received the benefit of a more powerful explosion without the expense of full royalty payment.

The torpedo company hired a legion of spies to report operators who patronized moonlighters, and the oil regions in the 1870's swarmed with investigators. It has been estimated that about two thousand prosecutions were threatened or instituted against operators who evaded the royalty fee by having charges set off at night. The courts always sustained the monopoly, and

[6] Peckham, *op. cit.*, 84.

the company made millions of dollars through its successful defense of its patent right and the royalty that it earned.

When nitroglycerin was first used in torpedoes, rarely were more than 10 quarts set off, but by the time the Bradford field was being developed in the late 1870's, it was not uncommon to use 40 to 120 quarts, in order to shatter a wider area of oil-bearing rock. Sometimes spectacular results were obtained by well owners who thought they might have drilled a dry hole.

Frank H. Taylor described such an incident:

On October twenty-seventh, 1884, those who stood at the brick schoolhouse and telegraph-offices in the Thorn Creek district and saw the Semple, Boyd & Armstrong No. 2 torpedoed, gazed upon the grandest scene ever witnessed in Oil-dom. When the shot took effect and the barren rock, as if smitten by the rod of Moses, poured forth its torrent of oil, it was such a magnificent and awful spectacle that no painter's brush or poet's pen could do it justice. Men familiar with the wonderful sights of the oil-country were struck dumb with astonishment, as they beheld the mighty display of Nature's forces. There was no sudden reaction after the torpedo was exploded. A column of water rose eight or ten feet and fell back again, some time elapsed before the force of the explosion emptied the hole and the burnt glycerine, mud and sand rushed up in the derrick in a black stream. The blackness gradually changed to yellow; then, with a mighty roar, the gas burst forth with a deafening noise, like the thunderbolt set free. For a moment the cloud of gas hid the derrick from sight and then, as this cleared away, a solid golden column half-a-foot in diameter shot from the derrick-floor eighty feet through the air, till it broke in fragments on the crown-pulley and fell in a shower of yellow rain for rods around. For over an hour that grand column of oil, rushing swifter than any torrent and straight as a mountain pine, united derrick-floor and top. In a few moments the ground around the derrick was covered inches deep with petroleum. The branches of the oak-trees were like huge yellow plumes and a stream as large as a man's body ran down the hill to the road. It filled the space beneath the small bridge and, continuing down the hill through the woods beyond, spread out upon the flats where the Johnson well is. In two hours these flats were covered with a flood of oil. The hill-side was as if a yellow freshet had passed over it. Heavy clouds of gas, almost obscuring the derrick, hung low in the woods, and still that mighty rush continued. Some of those who witnessed it estimated the well to be flowing five-hundred barrels per hour. Dams were built across the stream, that its production might be estimated; the dams overflowed and were swept away before they could be completed. People living along Thorn Creek packed up their household goods and fled to the hill-sides. The pump-station, a mile-and-a-half down the creek, had to extinguish its fires that night on account of gas. All fires around the district were put out. It was literally a flood of oil. It was estimated that the production was ten-thousand barrels the first twenty-four hours. The foreman, endeavoring to get the tools into the well, was overcome by the gas and fell under the bull-wheels. He was rescued immediately and medical aid summoned. He remained unconscious two hours, but subsequently recovered fully. Several men volunteered to undertake the job of shutting in the largest well ever struck in the oil-region. The packer for the oil-saver was tied on the bull-wheel shaft, the tools were placed over the hole and run in. But the pressure of the solid stream of oil against it prevented its going lower, even with the suspended weight of the two-thousand pound tools. One-thousand pounds additional weight were added before the cap was fitted and the well closed. A casing-connection and tubing-lines connected the well with a tank.[7]

By January 1, 1881, the Pennsylvania oil fields had produced an estimated total of 156,890,931 barrels of crude oil, more than 95 per cent of the nation's production.[8]

[7] Quoted in McLaurin, *op. cit.*, 338.   [8] Peckham, *op. cit.*, 149.

By the time the Bradford field was developed, geologists and men in the production business had learned to taste, smell, sample, and study particles brought up by the drill and sand pump, and to plot substratum formations. When Drake exultantly announced in August, 1859, he had "tapped the mine" after the bit dropped to a crevice seventy feet below the surface, early development continued in the valley of Oil Creek and its tributaries. With no guidance other than the earliest discovery, it was thought there was some connection or parallel between the fresh-water streams on the surface and the "oil veins" beneath. Hence, the first wells were drilled in the valley of Oil Creek or its tributaries from Titusville to Oil City, along French Creek from Union City to Meadville and Franklin, and on the Allegheny River at Tideoute.

Several failures to find oil along the stream beds, however, led to exploration on higher ground. This proved productive in many instances, particularly when the higher location was on a line between productive wells in the valley. Substance was given to the belief that producing wells could be developed regardless of the topography of the country.

The first wells were drilled until abandoned or an indication of oil appeared, without regard to the character of the strata pierced. Experience soon proved the sand rocks to be the source of petroleum, and along Oil Creek three oil-producing sands were discovered and classified; the third or deepest was the most productive.

As drilling progressed along the ridges, the drill passed through four, sometimes five, sands before reaching the depth of the first sand of Oil Creek. It was not unusual for errors to result in estimating depth to sands because of variance in surface elevations and dips in the oil-producing sands; sometimes, a thick stratum of sand at one location was split in two at another.

Drillers quickly learned to analyze sediment brought up by

the sand pump. Geologists, also, were interested in the borings, the depth of producing wells, and the extent or thickness of the producing sand. They could identify rock formations; they studied the dip, superposition, and extension of oil-producing sands. In 1865, for example, Professor J. P. Lesley was asked his opinion in regard to the probability of oil near Brady's Bend, on the Allegheny River below Oil Creek. He stated that the third sand of Oil Creek possibly extended that far but at a depth of 1,100 feet. Oil was struck at 1,120 feet.[9]

C. D. Angell developed a productive oil property in 1867 on Belle Island, in the Allegheny River, twenty-five miles below Oil City. He had drilled wells along Oil Creek and had become curious in regard to the location of the most productive wells in relation to one another. He theorized that oil lies in belts, in long and narrow areas having a general northeast and southwest extension, often no more than thirty rods in width but several miles in length. He had observed that a narrow belt extending from Scrubgrass, on the Allegheny River, to Petroleum Center, on Oil Creek, included many of the best wells in that region.

Angell knew that the sand rock was thickest and most productive along the axis of the belt, thinning out toward its borders. He concluded that the topography of the upper surface had no relation to the form, extent, or direction of the "belt." With these theories and facts in mind, Angell projected a similar belt, lying in a direction nearly parallel with the first, and extending from St. Petersburg in Clarion County, through Parker's Landing, to Bear Creek in Butler County. This proved to be the most productive area during the next decade.

In the 1860's, T. S. Hunt advanced the theory that oil comes from limestone at the base of the Devonian formation and was indigenous to those rocks with accumulations under the crowns of anticlines. A decade later geologists concluded that petroleum occurs in crevices only to a limited or unimportant extent; that

9 *Ibid.*, 12–13.

it was usually found saturating porous strata or overlying superficial gravel. In Canada and West Virginia, it was found beneath the crowns of anticlines; in Pennsylvania, it was found saturating porous portions of formations of various geological ages, between the Lower Devonian and the Upper Carboniferous, with oil sands generally sloping toward the southwest.

"The Bradford field in particular resembles a sheet of coarse grained sandstone 100 square miles in extent varying from twenty to eighty feet in thickness, lying with its southwestern edge deepest and submerged in salt water and its northeastern edge highest and filled with gas, under an extremely high pressure."[10]

By 1880, highly productive sands had been struck at depths varying from three hundred to two thousand feet, and cable tools were in use that weighed over two thousand pounds. In the early 1860's, when many of the shallow wells were "kicked down" by foot power and the spring-pole method, or small engines were used to furnish power, progress of two to ten feet a day was considered good. By the close of the 1870's, with the use of heavier tools dropping with every turn of the walking beam, heavier blows were struck and more rapid progress was made. Wells that fifteen years before had required thirty days to six months to complete could now be drilled in ten to thirty days' time.[11]

A string of drilling tools weighed about 2,100 pounds: the rope socket, about three and one-half feet long, 80 pounds; sucker bar, eighteen feet long, 540 pounds; jars, over seven feet in length, 320 pounds; a thirty-foot auger stem, 1,020 pounds; and the center bit, three and one-third feet long, 140 pounds.

Other tools weighed as follows: a temper screw, 145 pounds; eight-inch jars, 565 pounds; two eight-inch bits, 320 pounds; an eight-inch reamer, 180 pounds; two five-and-one-half-inch

10 *Ibid.,* 52.
11 *Ibid.,* 88.

bits, 280 pounds; a five-and-one-half-inch reamer, 140 pounds; a ring socket, 50 pounds; and two wrenches, 210 pounds.

The total cost of a set of tools was approximately seven hundred dollars; a driller's complete outfit, including cable, cost about nine hundred dollars.

Tools of various kinds for "fishing" other tools, broken or lost anywhere from one hundred to two thousand feet below the surface, had been developed, from the delicate grab, designed to pick up a small piece of valve leather or a broken sucker-rod rivet from the pump chamber to the ponderous string of "pole tools" containing tons of iron. The latter, at a depth of one thousand five hundred feet or more, could unscrew a set of "stuck tools" and bring them up piece by piece; or cut a thread on the broken end of a sinker bar or an auger stem, to which tools could be screwed fast.

After the drilling location was decided upon, the derrick built, tools, engine, and boiler in place, an eight-inch iron pipe was driven into the bedrock, to shut off ground waters. A smaller bit was then used and the hole contracted to five and five-eighths inches. A second tube, armed with a steel shoe, was then forced down inside the drivepipe and ground into the tapering drill hole to make a watertight joint. The eight-inch jars and tools were exchanged for five-inch tools, and the hole was drilled from that point, with the same diameter as the inside diameter of the casing. Only enough water was introduced to sand-pump properly. After a five-and-one-half-inch hole was drilled through the oil sand to the slate or shale beneath, a two-inch tubing was inserted, perforated in the oil sand, and a rubber packer inserted above the oil sand to seal the space between the tubing and the hole.

The cost of drilling to an average depth of 1,625 feet in 1880 was estimated to be approximately $3,400.[12] This included a complete rig with irons, $400; boiler and engine, $775; belt-

[12] Stowell, *op. cit.*, 196.

ing and connections, $125; casing, 500 feet at seventy-five cents a foot, $375; tubing, 1,700 feet at twenty-two cents a foot, $374; delivery of casing and tubing to well site, $35; packer, $15; drilling at sixty cents a foot, 1,625 feet, $975; tank, $160; and torpedo, $200.

In the first twenty-one years of the oil industry, the chief producing area was a broad elliptical belt extending beyond the western boundary of Pennsylvania into Ohio and West Virginia, and across Pennsylvania in a northeasterly direction into New York. By 1880, the Bradford field in Pennsylvania was producing more than 34,000 barrels daily. Production in the entire state was more than 65,000 barrels daily. During the ten-year period from 1872 to 1882, the average daily production in Pennsylvania and New York increased from 17,000 barrels to over 82,000, and the yearly production from 5,250,000 to 30,000,000 barrels. Annual production at this time in Ohio and West Virginia was less than 225,000 barrels and in California, about 150,000 barrels.

The first producing wells in the Bradford field were drilled in 1875, but it was not until two years later that its rich character became apparent. In the autumn of 1880, after four years of continuous drilling when the boundaries of the great reservoir had been determined, more than 9,000 wells were producing oil. Many of these were flowing wells; in defining the limitations of the pool, less than 5 per cent of the wells drilled were dry or failed to produce oil in paying quantities.

The flood of oil pouring from the Bradford field was more than trade conditions warranted and, as had so often happened since the Drake discovery well, supply far exceeded market demands. But production increased despite the oversupply and decline in price.

Producers had learned through years of experience in drilling for oil that, once the oil-bearing sand was tapped, because of the mobile quality of petroleum, the oil would flow to the well bore from its subterranean chambers. Knowledge of its migratory nature led to the practice of putting down as many wells as locations would permit, particularly along dividing lines of adjoining tracts. Usually land surrounding the discovery well was divided among a number of owners, each of whom hoped to tap the underground storehouse beneath his property before it was drained toward a rival location a few yards distant. Hence, there arose the practice of drilling too many wells when a new field was opened, and consequently, too much production.

Another cause of overproduction was the type of lease agreement in use. Landowners received a royalty payment varying from one-eighth to one-half of the production and, where chance of success seemed favorable, a bonus in money was added. Landowners and, beginning in 1860, stock companies that purchased the land quite often peddled leases for an acre or less of ground, accepting a money payment for the lease-right as well as reserving a royalty interest. By 1870, the lease always stipulated the number of wells to be drilled within a definite time limit and contained clauses of forfeiture for failure to carry out the agreement. These conditions impelled a producer to drill wells when the market price of oil did not warrant the expense. He did so rather than forfeit a lease on which he may have already made a valuable investment or payment, and because of a recurring belief that oil produced would rise in value.

The reason for a forfeiture provision in lease agreements was clearly expressed by an opinion delivered in 1875, by Daniel Agnew, chief justice of the Pennsylvania Supreme Court:

> The discovery of petroleum led to new forms of leasing lands. Its fugitive and wandering existence within the limits of a particular tract was uncertain, and assumed certainty only by actual development founded upon experiment. The surface required was often small compared with the results when attended with success, while these results led to great

speculation, by means of leases covering the lands of a neighborhood like a swarm of locusts. Hence it was found necessary to guard the rights of the landowners as well as public interest, by numerous covenants, some of the most stringent kind, to prevent their lands from being burdened by unexecuted and profitless leases, incompatible with the right of alienation and the use of the land. Without these guards, lands would be thatched over with oil-leases by sub-letting, and a farm riddled over with holes and bristled with derricks, or operations would be delayed so long as the speculator would find it hopeful or convenient to himself alone. Hence covenants become necessary to regulate the boring of wells, their numbers and time of succession, the period of commencement and of completion, and many other matters requiring special legislation. Prominent among these was the clause of forfeiture to compel performance and put an end to the lease in case of injurious delay, or a want of success.[13]

Whenever a new field was developed, the surface was "riddled over with holes and bristled with derricks," and adjoining property owners drilled protecting lines along property divisions. The courts early decided that "whatever gets into the well belongs to the owner of the well no matter where it comes from."[14] The theory of the court that oil underneath the surface of the earth was subject to capture, just as a wild animal is subject to capture and claim by the successful huntsman, made producers of oil practically helpless in regard to the number of wells drilled and the amount of oil produced. Waste of both gas and oil was greatly increased, of course, by the general recognition of the law of capture. The Pennsylvania Supreme Court recognized this law of the jungle when it decreed that "every landowner or his lessee may locate his wells wherever he pleases, regardless of the interests of others. He may distribute them over the whole farm,

---

13 *Brown* v. *Vandergrift*, 80 Penn. St. 142 (November, 1875).
14 *Kelly* v. *Ohio Oil Company*, 49 N.E. Report 401 (December, 1897).

or locate them only on one part of it. He may crowd the adjoining farms so as to enable him to draw the oil and gas from them. What, then, can the neighbor do? Nothing; only go and do likewise."[15]

Samuel H. Pettingill graphically illustrated in his book *Hot Oil* the effect that the interpretation of the rule of capture by early court decisions had upon the production of oil and gas:

1	2			3	
X	O	*Jones*	O	X	*Smith*
X	O				
X	O	W			
X	O		O'	X'	
X	O				
X	O		O	X	

Here we have three tracts of land. Number 1 has been subdivided into town lots and the lots sold to six different owners. Jones discovers oil in the center of his tract "W." Smith immediately drills a well on his tract as close to the boundary, and as near the proven well "W" as he can. We mark it "X'." "X" will drain oil from beneath Jones' land, so Jones drills an "offset," "O'" opposite "X'." Smith drills other wells along the line and Jones drills other offsets. Meantime the town lot owners get the fever and each of them drills, and Jones drills offsets opposite each of them. Soon there are 19 wells drilled where three, let us say, would have been, not only all that were necessary, but all that were desirable from the standpoint of getting out of the underground pool the largest per cent possible of its contents.

Result (as illustrated): sixteen unnecessary wells, costing

[15] *Westmorland and Cambria Natural Gas Company* v. *Dewitt et al.*, 130 Penn. St. 235 (November, 1899). The quotation is from a decision argued on October 18, 1906, in *Barnard* v. *Monongahela Natural Gas Company,* 216 Penn. St. 365 (October, 1906).

THE "RULE OF CAPTURE"

say $22,000 each (national average), or $352,000 of unnecessary expense which must, in the long run, be recovered
from the consumer in higher prices.

Result: nineteen wells going at full capacity, wasting gas
into the air, rapidly exhausting the reservoir energy underground, (like a locomotive blowing off steam) with an actual
recovery of, say, only 10 or 20 per cent of the underground
pool, instead of 50 or 60 per cent, before the energy is completely gone, and the oil either lost forever, or recoverable
only by pumping, re-pressuring, or mining—all at higher
prices to the consumer.

Result: The flood of oil from the new flush field temporarily oversupplies the market, and the capacity to store.
Price falls so that those who invested their money in the field,
as well as royalty owners, actually get fewer dollars and less
oil than if the field had been scientifically developed. In this
respect, it works out differently from the way it does with two
boys sucking lemonade. There, all the lemonade is recovered.
In an oil field, much is left in the ground, and that which is
brought to the surface commands a lower price. In the long
run, everybody loses.

Result: Persons operating in other fields hear of this new
flush field, abandon their wells and rush to the new field. Or
the price of oil brought about by the excess from the flush
field causes other fields to be "plugged." In either case, recoverable oil in the other fields is lost to the nation, either
absolutely, or through an increased cost when the oil fields
are reworked.[16]

Flush production in one area always endangered the life of
old wells in another. In 1860, with only two shallow wells competing with his, Colonel Drake thought there was overproduction; in 1861, with the first flowing wells on Oil Creek, many
shallow wells of small production were abandoned; the internal
revenue tax of one dollar per barrel of crude oil produced, en-

[16] *Hot Oil: The Problem of Petroleum*, 73–75.

99

acted in March, 1865, put hundreds of marginal producers out of business; a national depression, fluctuations in foreign consumption of American petroleum, and newly developed fields that increased the supply recurrently affected the price of oil and hastened the closing down of wells with small production.[17]

How widely the price of crude oil fluctuated on a day-to-day basis during the first decade of the oil industry has been described by McLaurin:

> Crude bought in September of 1862 at thirty cents a barrel sold in December at eleven dollars. John B. Smithman, Munhall's buyer, walked up the creek one morning to buy what he could at three dollars. A dispatch at Rouseville told him to pay four, if necessary, to secure what the firm desired. At Tarr farm another message quoted five dollars. By the time he reached Petroleum Center the price had reached six dollars and his last purchases that afternoon were at seven-fifty.[18]

Business was done on honor and agreements fulfilled to the letter, whether prices rose or fell. W. H. Abbott, a refiner at Titusville, paid $10.00 a barrel for crude oil in January, 1861, and only $1.25 in March. In October of the following year, the shipping firm of Howe and Nyce stored five hundred barrels of crude on the railroad platform of that city. This was sold to Abbott for $2.60 a barrel. In January, 1863, Abbott sold the same lot for $14.00 a barrel, and in March—it had never been moved from the platform—the oil brought $8.00 a barrel. Thirty days later Abbott bought it again at $3.00 a barrel and refined it.

Shortly after the Noble and Delamater well on Oil Creek gushed forth in May, 1863, Charles L. Wheeler, a buyer at Titusville, asked Orange Noble how much he would take for 30,000 barrels of the product. Noble asked $1.50 a barrel. Wheeler agreed to the price, but no written record was made of the transaction. Before the oil was delivered, the market price

[17] Giddens, in *The Birth of the Oil Industry,* 153–68, devotes a chapter to "The Depression Years."

[18] McLaurin, *op. cit.,* 264, 269.

100

had risen to $7.50 a barrel, but Noble delivered the quantity at the $1.50 price on the basis of the oral agreement made at their chance meeting some weeks before.

When a flood of oil was released by flowing wells along Oil Creek in 1861 and a limited demand depressed its price to ruinously low levels, William Coleman of the Columbia Oil Company proposed to make a lake of oil by excavating a pool of 100,000-barrel capacity. His suggestion was acted upon in the expectation that the supply would soon be exhausted, the price would rise to $10.00 a barrel, and the lake reserve would be worth $1,000,000. Evaporation and deterioration caused the loss of thousands of barrels of oil before the project was abandoned.

Ten years later a similar attempt was made on the McCray Farm, the most productive field in the lower Allegheny region at that time.

> Prices were hovering around three dollars, and McCray swore he would not sell under five dollars. He bought, hired, and built iron tankage until he had upward of 200,000 barrels. There was great loss from leakage and from evaporation and there were taxes, but McCray held on, refusing four dollars, $4.50, and even five dollars. Evil times came in the Oil Regions soon after and with them "dollar oil." McCray was finally obliged to sell his stocks at about $1.20 per barrel.[19]

Because of overproduction and limited demand for oil, producers and landowners formed the Oil Creek Association in 1861. Its primary purpose was to limit production from flowing wells in order to adjust the supply more nearly to the market demand. Before the plan could be fully tested, however, the price had advanced to $4.00 a barrel and the proration plan was abandoned as flowing wells depleted reservoir energy.[20]

[19] Ida M. Tarbell, *The History of the Standard Oil Company*, I, 32; also, Giddens, *The Birth of the Oil Industry*, 82, 183.
[20] McLaurin, *op. cit.*, 139; and Giddens, *The Birth of the Oil Industry*, 84.

In May, 1865, landowners and producers of the Tarr Farm met and successfully inaugurated the practices of casing wells to prevent water seepage into the oil sands, and the plugging of abandoned wells. Producers along Oil Creek noted benefits from the experiment and began a campaign to make the plugging of abandoned wells obligatory. Jonathan Watson reported that water flooded one of his wells when the seed bag was removed from another well one-half mile distant. In another instance, when Watson was testing the porosity of oil sand at one of his locations, red paint was put into one well and pumped out at another several hundred yards away. Although several attempts were made to enact state legislation to make the plugging of abandoned wells mandatory, this was not accomplished until 1878. The act provided for a fine unless the well was filled with sand to exclude water from oil-bearing rock and to prevent the flow of oil and gas into fresh water.[21]

Prior to formal organization of oil exchanges, deals between producers, refiners, buyers, shippers, and speculators were made through personal contact at the well, boardinghouses, and hotels, and by chance meetings on the streets of Oil City, Titusville, and other points of oil activity. Purchases for Pittsburgh, Baltimore, and Philadelphia refiners started brokerage in 1868, on a commission of ten cents a barrel from buyers and five from sellers. A principal trading center, locally known as the "curbstone exchange," was the board sidewalk on Center Street in Oil City, in front of the shipping firm of Lockhart, Frew and Company. After Titusville and Oil City were connected by rail in 1866, a special coach was attached in which producers, dealers, and speculators transacted business over the seventeen-mile stretch during the three-hour run, with stops at settlements along the way. Deals were made for "spot," "regular," and "future" oil. "Spot

---

[21] "An Act Requiring Owners and Operators of Oil Lands to Plug Their Wells," approved on May 16, 1878, in *Laws of the General Assembly of the State of Pennsylvania,* passed at the session of 1878, 56–57.

oil" was to be moved and paid for at once, "regular" allowed the buyer ten days for shipment, and "future" stipulated an agreed-upon date for delivery.

Finally, producers, refiners, dealers, and brokers met in January, 1871, at Titusville and organized an oil exchange. Another was soon organized at Franklin, then Oil City, followed by others in Parker, Bradford, Pittsburgh, Philadelphia, and New York. Typical of the exchanges is the description of the one at Oil City:

> In February of 1874 the exchange was reorganized, with George V. Forman as president, and occupied quarters in the Collins House for four years. Gradually rules were adopted and methods introduced that brought about the system later in vogue. In April of 1878 the formal opening of the splendid Oil Exchange Building took place. The structure contained offices, committee-rooms, telegraph-lines, reading-rooms and all conveniences for its four hundred members. H. L. Foster was president term after term. H. L. McCance, secretary for years, was a first-class artist, with a skill for caricature worthy of Thomas Nast. Some of the most striking cartoons pertaining to oil were the work of his ready pencil. F. W. Mitchell & Co. inaugurated the advancing of money on certificates, their bank's transactions in this line ranging from one to four million dollars a day. The application of the clearing-house system in 1882 simplified the routine and facilitated deliveries. The volume of business was immense, the clearances often amounting to ten or fifteen million barrels a day. Only the New York and San Francisco stock exchanges surpassed it. If speculation were piety, everybody who inhaled the air of Oil City would have been saved and the devil might have put up his shutters. During rapid fluctuations the galleries would be packed with men and women who had "taken a flyer" and watched the antics of the bulls and bears intently. Fortunes were gained and lost.[22]

22 McLaurin, *op. cit.*, 282.

103

7

GROWING PAINS OF THE INDUSTRY

FROM 1872 TO 1882, the average daily production of crude oil from fields extending across Pennsylvania into the state of New York increased from 17,000 barrels to 82,000 barrels, and the yearly production from 5,205,234 to 30,053,500 barrels. Ohio and West Virginia produced 224,312 barrels in 1880, and California produced 180,000 barrels.[1] This was a period of rapid development and improvement in the transportation of crude oil and its products, and it marked a vigorous struggle by producers against the monopolistic tendencies of the Standard Oil Company, which was formed in 1870 to refine petroleum.

When Drake "tapped the mine," the world was seeking a better illuminant than was afforded by the open fireplace, the tallow candle, the pine knot, or a lamp filled with sperm or whale oil. The lessening supply of sperm and whale oils and their advance in price led to attempts to find or produce substitutes, and as early as 1840, coal oil was being manufactured in limited quantities in France. Soon thereafter, E. W. Binney and Company of Bathgate, Scotland, produced oil from coal and began to supply a demand for the product in Great Britain. James Young, a member of the company, introduced the process into the United States, and parties in New Bedford, Massachusetts, who had been engaged in processing sperm and whale oil, began the manufacture of coal oil from mineral imported from Scot-

[1] Stowell, *op. cit.*, 189–201.

land. Rich cannel coals of western Virginia and Kentucky soon attracted attention, and plants for the distillation of coal oil were operated in those states, as well as in Ohio and Pennsylvania.

In 1853, Luther and William Atwood established the United States Chemical Manufacturing Company at Waltham, Massachusetts, and began the manufacture of "coup oil" from asphalt or coal tar. This product, when mixed with cheap animal and vegetable oils, was used to lubricate machinery. By 1858, other plants were established by the firm in Portland, Maine, in Boston, and in Brooklyn, and they had begun experiments in refining petroleum. Samuel Downer, a Boston manufacturer of sperm-oil candles, purchased the company in 1854, but the Atwoods remained with the firm, as likewise did Joshua Merrill, their associate.[2]

Before the drilling of the Drake discovery well, Samuel Kier, the firm of MacKeown and Finley of Pittsburgh, and A. C. Ferris of New York were refining crude oil for illuminating purposes on a small scale, in stills of five-barrel capacity, or less, each twenty-four hours. Although the flame was more brilliant than that given off by the coal product, the odor was offensive. With the increased quantities of crude oil from wells, a coal-oil manufacturer who visited the oil fields in 1860 predicted, "If this business succeeds, mine is ruined."[3] Manufacturers of coal oil either went out of business or modified their plants so as to be able to refine petroleum.

Downer converted his plants on the seacoast from coal-oil distilleries to petroleum refineries. In 1862, he came to Corry, fourteen miles from Titusville, at the junction of the Philadelphia

[2] *Ibid.,* 187. Also, Peckham, *op. cit.,* 9–10; Giddens, *The Birth of the Oil Industry,* 19–22, 91–93. Abraham Gesner, a Canadian geologist, was the first American to produce an illuminating oil (in 1846). He named the product "keroselain"—later shortened to "kerosene"—a name applied to all mineral illuminating oils.

[3] Quoted in Giddens, *The Birth of the Oil Industry,* 95.

and Erie and the Atlantic and Great Western railroads, and constructed a well-equipped plant near the source of supply. The Atwoods and Merrill perfected the refining process to produce a better illuminant, and in 1869, Merrill discovered a process by which a deodorized lubricating oil could be made.

There existed a world-wide demand for a cheap illuminant. Kerosene from petroleum and improved lamps to burn the product opened a vast and profitable foreign market. Fortunately for the growing new industry, a resident of the Venango oil region, Dr. A. W. Crawford, was appointed United States consul to Antwerp, Belgium, in 1861, where he served until 1866. Shortly after his arrival he made the following report:

> I deem it a part of my duty to intimate that some of the numerous oil springs of the Alleghany and Kanawha valleys might profitably let their lights shine in this direction. The distillation of coal oil has but recently been carried on in Belgium. Nor is the Belgium coal largely imbued with gas; consequently, linseed oil is as yet the chief illumination.[4]

McLaurin has described the beginnings of foreign trade for the industry:

> Dr. A. W. Crawford, who served three terms in the Legislature, was appointed consul to Antwerp by President Lincoln, in 1861. At the time he reached Antwerp a cheap illuminant was unknown on the continent. Gas was used in the cities, but the people of Antwerp depended mainly upon rape-seed oil. Only wealthy people could afford it and the poorer folks went to bed in the dark. From Antwerp to Brussels the country was shrouded in gloom at night. Not a light could be seen outside the towns, in the most populous section on earth. A few gallons of American refined oil had appeared in Antwerp

---

[4] "U. S. Consul Suggests the Marketing of Petroleum in Antwerp," Document No. 106, in *Pennsylvania Petroleum,* 217.

previous to Dr. Crawford's arrival. It was regarded as an object of curiosity. A leading firm inquired about this new American product and Dr. Crawford was the man who could give the information. He was from the very part of the country where the new illuminant was produced. The upshot of the matter was that Dr. Crawford put the firm in communication with American shippers, which led to an order of forty barrels by Aug. Schmitz and Son, Antwerp dealers. The article had tremendous prejudice to overcome, but the exporters succeeded in finally disposing of their stock. It yielded them a net return of forty francs. The oil won its way and from the humble beginning of forty barrels in 1861, the following year witnessed a demand for fifteen-hundred-thousand gallons. By 1863 it had come largely into use and since that time it has become a staple article of commerce. Dr. Crawford served as consul at Antwerp until 1866, when he returned home and began a successful career as an oil-producer. It was fortunate that Col. Drake chanced upon the shallowest spot in the oil regions where petroleum has ever been found, when he located the first well, and equally lucky that a practical oil man represented the United States at Antwerp. Had Drake chanced upon a dry-hole and some other man been consul at Antwerp, oil-developments might have been retarded for years.[5]

In 1861, 20,000 barrels of American crude oil were exported to Great Britain; by the end of the Civil War petroleum ranked sixth in our exports, and by the end of the decade, more than 100,000,000 gallons were exported annually. Northern Europe yearly quickened its demand for the illuminant. The market had become world wide, with Antwerp the great entrepôt for European trade, and dealers in oil at the exchanges awaited eagerly the daily two o'clock Antwerp quotations.[6]

It will be recalled that in 1859 officers of the Seneca Oil

[5] McLaurin, op. cit., 212–13.
[6] Titusville Morning Herald, July 11, 1870.

Company sent a sample of petroleum to Havre, France, and had it analyzed by A. Gelee, a French chemist. He reported that it was suitable for lubrication and for making dyes, and that, with proper distillation, it made an excellent illuminant.

News of Drake's discovery, too, excited interest in England, where shale oil was being distilled for illuminating purposes. Drake received a letter early the next year from F. Streng and Company of Manchester, in which mention was made of his "extraordinary discovery of oil possessing superior qualities." The company requested Drake to set up a general agency for the Continent and Great Britain.

The first ship whose principal cargo was oil to clear an American harbor was the brig *Elizabeth Watts*. It was loaded with barrels of oil at Philadelphia in late November, 1861, and headed for London, the shipping rate being one dollar a barrel plus a 5 per cent commission to the owners of the vessel.

> Very little is known about the trip. It took several weeks to load the little oil sailor, and when she was ready for sea the skipper could not get together a crew to work above a cargo of oil. Failing to engage sailors in the regular way, men were got aboard under the influence of liquor, and she sailed down the Delaware River with a drunken crew. She got safely across the Atlantic, and landed her cargo in good condition at a London wharf.[7]

Other sailing vessels were adapted to the trade; during 1864, despite the threat of disaster at sea from fire or the Confederate blockade, 32,000,000 gallons of crude oil were shipped to Europe. By the end of the decade, refined as well as crude oil was being shipped in sailing vessels, and experiments proved the

[7] J. D. Henry, *Thirty-five Years of Oil Transport: The Evolution of the Tank Steamer*, 5. *The Derrick's Hand-Book*, I, 24, mentions that the first shipment of crude oil to the Netherlands was made in October, 1861.

practicability of shipping oil in bulk. Vessels were fitted with iron tanks to carry oil in separate compartments.

Owners of steamship lines, however, were more hesitant about converting to carriers of oil in bulk. A writer in 1879 reported:

> The report from Philadelphia that the steamer Vaderland of the Red Star line has been purchased by a number of capitalists for the purpose of transporting oil in bulk has attracted considerable attention at the various oil exchanges. The transportation of oil in bulk is not entirely an experiment. A number of sailing vessels have already been fitted out for this purpose and have, to a certain extent, demonstrated the practicability of the idea. This is the first time, however, that a steamer has been constructed solely with the view of transporting safely large quantities of petroleum in bulk.
>
> Inquiry among petroleum and shipping merchants in the city elicited the general opinion that the idea is not considered practicable. Said one well-known oil representative: "It is my opinion that the system will not work. It has been tried three times on sailing vessels during the past eight years and each time the vessel was lost. The captain of one of them, who was saved from the wreck of his vessel, said to me that the difficulty was that the oil seemed to move faster than water, and in rough weather, when the vessel was pitched forward, the oil would rush down and force the vessel into the waves much the same as improperly stored bulk grain does sometimes in stormy weather. It may be that by storing the oil in small compartments it could be transported with safety but I doubt it."
>
> Messers. Slocovich and Company, the well-known shipping merchants, stated that about eight years ago one of their vessels was fitted up with tanks for transporting oil in bulk. She proceeded on her journey and was never heard from. Her loss was undoubtedly due to her mode of carrying petroleum. Another shipping merchant stated that he believed the idea

to be impracticable. It might be possible to make the tanks strong enough to prevent the escape of the vapor of the oil, but all previous experiments had proven failures, and there was no reason to suppose that this would succeed. An experiment to transport molasses in bulk had been tried within two or three years, and two vessels were fitted for the purpose to run between Cuba and Boston. The experiment, however, proved a failure, and the project had been abandoned. . . .

The oil in bulk movement does not meet with favor among practical exporters. They say that it cannot be carried out successfully. It would seem, however, that oil might be transported in vessels in that way as well as grain, and the day will no doubt come when a means to that end will be devised.[8]

That day came sooner than expected. By 1885, more than one thousand vessels were engaged in the American export oil trade and the movement was underway to handle it in bulk with a capacity of 2,500 to 14,000 barrels a voyage. By the turn of the century, ships of the *Narragansett* class, divided into twenty-seven oil-tight compartments, were carrying up to 75,000 barrels a voyage. They could discharge the cargo and be ready to sail again within forty-eight hours. A tank steamer carrying 14,000 barrels of oil in bulk could load or unload in ten hours; if the oil was barreled, it took a week, and gangs of labor.[9]

The refining industry was heavily concentrated in the oil-producing region along Oil Creek and the Allegheny River during the first decade of activity; other principal centers were Cleveland, Pittsburgh, and New York. Competition and marketing problems had eliminated those of smaller capacity, and as the decade closed, a struggle was underway for further consolidation and control of the industry.

In October, 1910, an attorney for the Standard Oil Company of New Jersey expressed to the Supreme Court tribunal the eco-

---

[8] Peckham, *op. cit.*, 101.
[9] J. D. Henry, *op. cit.*, 31, 74.

nomic philosophy that led to its dominating place in the industry: "The elimination of competition is not a restraint of trade, but is merely an incidental effect of the exercise of the fundamental civil right to buy and sell property freely."[10]

Methods employed in eliminating competition, however, were questionable, and evidence presented to the Supreme Court indicated that the company had "obtained rebates and discriminatory rates in transportation of their product as against their competitors, and engaged in unfair and oppressive methods of competition, thereby destroying the smaller manufacturers and dealers throughout the country."[11]

The Standard Oil Company controlled 90 per cent of the shipping, refining, and marketing of petroleum and petroleum products by 1882, when it added production to its integrated operations. Leaders of the company foresaw in its formative years that consolidation and control of pipe lines and refineries would give them control of the industry. How well it succeeded was expressed by Mr. Justice Holmes when he delivered the opinion of the Supreme Court in the pipe-line cases, in October, 1913:

Availing itself of its monopoly of the means of transportation, the Standard Oil Company refused through its subordinates to carry any oil unless the same was sold to it or to them and through them to it on terms more or less dictated by itself. In this way it made itself master of the fields without the necessity of owning them, and carried across half the continent a great subject of international commerce coming from many owners but, by the duress of which the Standard Oil Company was master, carrying it all as its own.[12]

The Standard Oil Company was the outgrowth of a small

[10] 221 *United States Reports,* 7.
[11] *Ibid.,* 11.
[12] 234 *ibid.,* 559.

Cleveland refining company, backed by John D. Rockefeller and associates, which was incorporated in 1870. Cleveland was the chief competitor of the Oil Creek refiners, whose geographical proximity to production was offset by the larger city's location on the shore of Lake Erie. Cheap water transportation was afforded Cleveland by lake and canal. Two trunk railroad lines connected it with New York and the Atlantic coast; by rail and water, too, it was favorably located to control the western trade.

As the world demand for petroleum increased, competition for its transportation and manufacture also became keener. Marginal operators in the refining industry—those with small capacity—were placed at a disadvantage in competing for markets with those of larger capacity, and the geographical advantage of refineries located in the oil regions was offset by the rebate system that had become common by 1870. The theory developed and was accepted by the railroads that the biggest shippers should have the best rates, although railroads, as common carriers, had no right to discriminate in rates between customers.

In hopes of eliminating economic waste and instability in the competitive manufacturing process, the South Improvement Company was formed by leading refiners in Philadelphia, Pittsburgh, Cleveland, and New York. Contracts were then entered into with the railroad carriers, who agreed to a rebate system and foresaw a year-round stabilization of traffic in oil. Thomas A. Scott of the Pennsylvania Railroad, which controlled the Empire Transportation Company, its pipe lines, and tank cars, at first demurred to the agreement unless consideration was given to the producers. W. G. Worden of the South Improvement Company strongly objected to including producers in the scheme: "The interests of the producers were in one sense antagonistic to ours: one as the seller and the other as the buyer. We held in argument that the producers were abundantly able to take care of their own branch of the business if they took care of the quantity produced."[13]

112

Quickly, pledged to the utmost secrecy, members of the South Improvement Company in January, 1872, approached competitors and advised them to sell to the combine or be crushed. Frank Rockefeller, brother of John D., gave evidence of this scheme before a Congressional investigating committee four years later:

"We had in Cleveland at one time about thirty establishments, but the South Improvement Company was formed, and the Cleveland Companies were told that if they didn't sell their property to them it would be valueless, that there was a combination of railroad and oil men, that they would buy all they could, and that all they didn't buy would be totally valueless, because they would be unable to compete with the South Improvement Company, and the result was that out of thirty there were only four or five that didn't sell."

"From whom was that information received?" asked the examiner.

"From the officers of the Standard Oil Company. They made no bones about it at all. They said: 'If you don't sell your property to us it will be valueless, because we have got advantages with the railroads.' "

"Have you heard those gentlemen say what you have stated?" Frank Rockefeller was asked.

"I have heard Rockefeller and Flagler say so," he answered.[14]

Advantages in shipping rates accorded members of the South Improvement Company may be illustrated by citing rates to and from Cleveland. A charge of 80 cents a barrel for crude oil shipped from the oil regions to Cleveland was posted by the railroads. The South Improvement Company received not only a rebate of 40 cents a barrel, but a drawback from the railroads of 40 cents on each 80 cents paid the carrier by the independent refiner. The rate on crude oil shipped from the oil region to

13 Tarbell, *op. cit.*, I, 60.
14 *Ibid.*, 64.

New York was posted as $2.56, but members of the South Improvement Company received a rebate of $1.06, as well as cash in like amount on each barrel of crude shipped by parties outside the combination. Similarly, the cost of shipping the refined product from Cleveland to New York was posted at $2.00 a barrel; the combination paid $1.50, and received a 50-cent cash drawback per barrel on all refined products shipped by competitors.

When the new shipping rates were posted on February 26, wild excitement pervaded the oil regions. A chronicler of this period has written:

> On February 27, the streets of Titusville, Pennsylvania, were black with demonstrating oil-diggers. The Oil City *Derrick* had made known the names of the South Improvement ringleaders in a blacklist. Mass meetings, parades, speeches exhorting to burn the enemy refiners' oil, to tap the enemy tanks, to lynch the "conspirators." A secret association of the independent oil men was formed at once and bound its members by fiery oaths and ritual to "unite against the common enemy," to stop all oil production, to sell no more to the refiners who were members of the Combination or Anaconda. Petitions were addressed to the Pennsylvania legislature and to Washington. Among the practical measures urged by the marchers and demonstrators was the building of an independent railroad freight line, since from the existing railroads the oil men "expected only robbery." It was further advocated that an independent pipe line be built by government subsidy.[15]

As a result of the indignation meetings, there was formed the Petroleum Producers' Union, which was pledged to support refiners in the oil region and refiners in distant cities designated by the union's executive committee. A pledge was exacted from the

[15] Matthew Josephson, *The Robber Barons,* 160.

114

producers to sell their oil only through or with the consent of the committee of the Petroleum Producers' Union. A tax of one cent a barrel was exacted to pay the expenses of the committee, which was formed by representatives of sixteen districts into which the oil region was divided.[16]

Immediately the producers began their campaign to supply only refiners in the oil region and others known to be nonmembers of the combine. A boycott was effected against buyers of the South Improvement Company; more important, public opinion favored the producers. In April, a Congressional committee held hearings in regard to the combination, after which the chairman declared: "Your success meant the destruction of every refiner who refused for any reason to join your company, or whom you did not care to have in, and it put the producers entirely in your power. It would make a monopoly such as no set of men are fit to handle."[17]

Within a month after the discriminatory railroad rates had inadvertently become public knowledge, they were rescinded, and the carriers agreed to equitable rates. About the same time, the Pennsylvania Legislature repealed the charter of the South Improvement Company.

Methods of consolidation practiced by the South Improvement Company were followed by the Standard Oil Company, and producers in the oil regions were unable to keep a solid bloc of opposition. The company continued to absorb independent refiners, to receive rebates and drawbacks from railroads, to extend its terminal and marketing facilities, and to gain control of the principal gathering lines of the oil region.

Producers were able to muster resentment—but little else—against a system that refused transportation of crude from their storage tanks to railroad terminals unless for immediate ship-

[16] "Organization of the Petroleum Producers' Union of 1872," Document No. 8, in Tarbell, *op. cit.*, 292–94.
[17] Quoted in *ibid.*, 80.

ment at less than the market price. They resented the impediments placed in the way by railroad lines in furnishing tank cars for the shipment of crude oil to independent refiners. They did nothing, however, toward stemming the great flood of overproduction from the Bradford field in 1878.

The Petroleum Producers' Union sent a committee of twenty-five men to Harrisburg to see Governor John F. Hartranft. They presented an appeal to him which listed the rise of monopolistic practices in the oil industry since 1871, the consolidation of pipe lines, and the rate discriminations of railroads which destroyed the geographical advantage of refineries located in the oil region; they also pointed out current abuses affecting producers. They demanded that the United Pipe Lines be made to perform its duty as a public carrier and that railroads cease discrimination through rebates and allocation of tank cars.

The Union summarized its grievances as follows:

> By the theory of law, corporations derive their power from the people of the commonwealth in General Assembly convened; they have no powers not delegated to them by the people. . . . The railroad and pipe-line companies are common carriers of freight for all persons, are bound to receive it when offered at convenient and usual places, and to transport it for all, for reasonable compensation, without unreasonable compensation in favor of any. These are but simple statements of well established legal principles, never doubted in any court, but affirmed by every tribunal that has ever considered them. Yet the people who granted these special privileges are now upon the defensive, their rights denied by these corporations, and they are challenged to enter the courts to establish them, while in the meantime they are inoperative to the irreparable injury of their business. They have yielded to the railroads that they have created a part of their sovereignty, and given them the right to take private property for public use, but restricting such taking, strictly to such use. Yet where the

116

narrow strip of land used as a railway roadbed runs through valuable oil lands, this combination is strong enough to demand from the railways its transfer to them, that they may and do thereon sink their own oil wells, and thereby drain the oil from the adjoining lands whose owners gave the strip for public use by a railroad.

The owners of lands along the line of the Allegheny Valley Railroad, producing petroleum from those lands, with their own pipe-line running to their own shipping racks by the side tracks of that railroad, are unable to obtain cars in which to load their product for transportation, at any rate of freight, while their tanks overflow. Shippers of petroleum are refused cars, or are promised them, only to find the promises broken, and their contracts rendered impossible of fulfillment, while the monopoly demands and is given all the cars belonging to the railroads, it permitting its own private cars to meantime stand idle, so that the railroads may assert its inability to accommodate all.

Owners of tanks connected with the monopoly pipe-lines, with ample storage therein for their own product, are refused transportation from their own wells upon the ground that "their tanks are full," a barefaced and daily demonstrated falsehood. Other producers of petroleum are refused transportation by the pipe-lines, on the plea of want of capacity to carry, and at the same time are informed that their oil will be carried if they will sell it to the ring, "immediate shipment."

If the applicant's tanks are overflowing, or if he needs money and complies with their terms, he is offered a price from two and a half to twenty-five cents below the market value. If he accepts and sells a fixed amount of his oil, the pipe-line removes all but five or ten barrels, delays for days and weeks to take the remainder, and refuses to pay for any until all is taken. This is known as the "immediate shipment swindle."

These corporations have made themselves the interested

117

tools of a monopoly that has become the buyer, the carrier, the manufacturer, and the seller of this product of immense value. It needs no argument or illustration to convince that in such a position this foreign corporation is in direct antagonism to the producer, the laborer, and the consumer.

The South Improvement conspiracy embraced in its scheme the ownership of the oil producing territory, wells and machinery. If the present course of its successor cannot be stayed, it is merely a question of time when the ownership of the entire oil production will fall into its hands through the impoverishment of thousands of our citizens and their inability to contend longer.

That monopolies are dangerous to free institutions is a political maxim so old as to have lost its force by irrelevant repetition, but if anything were needed to awaken the public sense to its truth, the immediate effect of this giant combination is before us. Throughout the Oil Region, as wherever it does business, it now has its own acid works, glue factories, hardware stores and barrel works. We have seen that it is master of the railroads, and owns and controls all the refineries, all the pipe-lines. All these enumerated industries controlled by them employ large numbers of laborers dependent for the support of themselves and their families upon the daily labor given or withheld by this powerful conspirator. At the flash of the telegraphic message from Cleveland, Ohio, hundreds of men have been thrown out of employment on a few hours' notice and kept for weeks in a state of semi-starvation and justifiable discontent, deceived meanwhile with delusive promises of work, until the autocrat of a foreign corporation, maintained and upheld by the chief among Pennsylvania corporations, gives leave from within the borders of a foreign state for the Pennsylvania laborer to earn his bread.

Along the valley of Oil Creek and the Allegheny Valley, where a few years since the smoke of busy refineries and their attendant industries darkened the air, piles of rusted iron and heaps of demolished brick work mark the results of the conspiracy; where a few years since busy men crowded to and fro in the pursuit of lawful trade in a great staple, there is now silence and emptiness. The producer, once surrounded with competitive buyers of his product, now goes with crowds of his fellow victims to wait his turn for leave to sell at a dictated price to a single agent of a single purchaser.

So far as this public wrong is within the scope of Executive interference, we ask that immediate steps be taken to enforce by legislative enactment the wise provisions of our State Constitution, and by such legal processes as are necessary, compel obedience to law and the performance by chartered companies of their public duties.[18]

Although practices by the Standard Oil Company did not violate the principles of morality invoked by contemporary big business, they were offensive to the principles of competition in industry: they placed the producers at a price disadvantage and effectively controlled the marketing of oil and refined products. In two years' time the Producers' Union, discouraged by delays in seeking redress through court action and by a change in state administration which showed little interest in their problems, deserted by men who for eight years had fought for the principle of independence, had only one hope of freedom left: the construction of a seaboard pipe line, which would free them from the railway monopoly and furnish an outlet to independent refineries and the foreign market.

During the first ten years following the introduction of the use of pipe lines in 1865, a number of companies were organized for the running and storing of crude oil. Gathering lines of companies served a restricted territory, and there arose the practice of giving producers a certificate or receipt for their oil which was redeemable in oil at any time upon payment of storage

[18] "Producers' Appeal of 1878 to Governor John F. Hartranft, of Pennsylvania," from *House Report 3112* (1888), 351–56, listed as Document No. 32, in *ibid.*, 381–90.

charges. In the beginning, the certificates were redeemable only in the district where the pipe-line company operated and were not negotiable.

By the mid-seventies, pipe-line carriers had systematized their operations. When a producing well was struck, a feeder line was run to the well free of charge, and the amount of the oil received in the line was ascertained and placed to the credit of the well owner. Three per cent was deducted to cover loss in transportation. Oil was held by the gatherer for the owner, or it could be transferred from one account to another upon written order. Upon signature of the owner for all or any part of his credit balance, the order was marked "accepted" by the company. This order or certificate passed from hand to hand like a certified check. Certificates were usually issued per 1,000 barrels of oil and were traded as any negotiable paper. When an oil owner, or owner of certificates, wanted to use his oil, he paid a storage charge of twenty cents a barrel, and the oil was delivered out of the lines of the company. A nominal assessment against holders of certificates or oil was charged by the company when oil was destroyed by fire or other accidents. The assessment, known to the trade as the "general average," was made on all oil in the care of the company. No matter how great the loss in one locality, the general assessment of all owners of oil throughout the system was nominal, and it was generally accepted as a cheap form of insurance.[19]

Typical of this mutual insurance program is the following notice posted by William T. Scheide, general manager of the United Pipe-Lines, on August 30, 1880. "The patrons of the United Pipe-Lines are hereby notified that all credit balances upon the books of the United Pipe-Lines at the close of business August 28, and all outstanding acceptances issued on or before that date, are subject to an assessment of twenty-one one-hundredths (21/100) of 1 per cent in pipeage paid oil, on ac-

[19] S. H. Stowell, *op. cit.*, 205.

count of loss by fire, on August 28, 1880 of tank United register No. 738, located at Babcock, on the Erie railroad, McKean County, Pennsylvania."[20]

In October, 1877, the Standard Oil Company purchased from the Empire Transportation Company, an affiliate of the Pennsylvania Railroad Company, its pipe lines, its refineries in New York and Philadelphia, and its oil-tank cars. Two months later the Columbia Conduit Pipe Line, extending from the lower oil region of western Pennsylvania to Pittsburgh, was absorbed by the United Pipe-Line Company. These acquisitions gave Standard complete control of the gathering lines—from well to railroad—in the oil regions.

In 1878, the Petroleum Producers' Union became interested in a pipe line extended to the tidewater region. A seaboard line would free them from discriminatory rates imposed by the railroads and afford an outlet to foreign markets. In November, 1878, a limited partnership, the Tidewater Pipe Company, was organized to carry out the scheme. The open rate for shipping oil by rail to the seacoast was $1.25 to $1.40 a barrel; General Herman Haupt, the engineer for the project, estimated pipes could carry the oil for 16 2/3 cents a barrel. This competition foredoomed railroads as monopolistic carriers of crude oil and threatened to jeopardize the favored position attained by the Standard Oil Company.

By the following May, 109 miles of the six-inch line were completed; up to that time, oil had never been pumped more than 30 miles. A series of powerful pumps were installed, capable of lifting the oil 700 feet; a total elevation of 2,600 feet had to be overcome before the gravity flow could carry the oil to the seaboard.

The success of the Tidewater venture was quickly challenged by Standard. The National Transit Company was organized in 1881, and pipes were laid to Cleveland, Buffalo, Pittsburgh,

[20] Peckham, *op. cit.*, 98.

Philadelphia, and Bayonne, New Jersey. Minority stock in the Tidewater Pipe Company was purchased by Standard, and in 1883, an agreement was effected whereby 11.5 per cent of the carrying trade to the Eastern seaboard was assigned to Tidewater, with Standard taking the remainder.

With the development and extension in the 1880's of prolific oil fields in Ohio, West Virginia, and Pennsylvania, and attendant increases in daily production, the Standard Oil Company bought producing properties. This step was taken to counteract a movement by the Producers' Protective Association, begun in 1887, to organize a company for the establishment of refineries and an interstate pipe line. Standard thus became a completely integrated concern—with producing properties, storage and transportation facilities, refineries, and marketing outlets.

Despite inroads made into the Producers' Protective Association when some of its principal members sold their producing properties to the Standard Oil Company, the organization persisted in its efforts to free itself from dominant control. Members of the association combined to form the Producers' Oil Company, the Producers' and Refiners' Company, and the United States Pipe-Line Company, which was projected as an interstate carrier to the eastern seaboard. In November, 1895, the independents incorporated in New Jersey a marketing concern called the Pure Oil Company. Five years later majority stock of the producing, carrying, and refining companies was transferred to this company; most of the nine hundred stockholders were men associated with production activities.

As the kerosene age drew to a close, men who for thirty years had been active in the oil industry, who had witnessed improvements in drilling techniques, in oil field machinery, in railroad tank cars, in pipe lines, in refining processes, and in marketing facilities, and who had watched the ruthless though efficient operations of the Standard Oil Company, at last realized the industry had grown too big to be completely dominated by one concern. The formation and survival of the Pure Oil Company in the face of intense competition, by adapting the integration policies so successfully maintained by its principal rival, presaged a rebirth of competition in the industry. As the century drew to a close, the Pure Oil Company had fourteen refineries—the majority located in the oil region—1,500 miles of crude oil pipe lines, and 400 miles of line to carry refined products, tank steamers, oil barges, and domestic and foreign market outlets. In May, 1901, the first refined oil piped from the oil region to the seaboard was transported in this company's lines. A new era of competition was approaching as vaster and more important industrial uses of oil came into being.

8

EXPANSION OF

THE OIL INDUSTRY, 1900-50

THE development of the internal-combustion engine in the 1890's, the advent of the automobile, and its rapid ascendancy in transportation set off the phenomenal growth of the petroleum industry during the first half of the twentieth century.[1] Commonplace instances come to mind: the development of the automobile for family and business use; buses, vans, and trucks that revolutionized the transportation industry; tractors and motor-driven farm equipment; the conversion of locomotives and seagoing vessels to the use of petroleum fuels for power; factory and industrial uses of residual and fuel oils; the rapid expansion of aviation; and the dominant role played by petroleum in two global wars.

There are many ways of stating the remarkable rise of the petroleum industry during this period. For example, in 1955, there were twelve industrial corporations in the United States with a net worth of more than one billion dollars: Standard Oil (New Jersey), the General Motors Corporation, United States Steel, the Du Pont Company, the Ford Motor Company, Socony Mobil Oil, Standard of Indiana, the Texas Company, Gulf, Standard of California, Bethlehem Steel, and the General Electric Company. It is significant that among the first ten, six are oil companies and two of the remainder represent the automobile industry. Also, in 1955, twenty-six companies of the country

[1] A. I. Levorson, *Geology of Petroleum*, 4.

124

had net sales in excess of one billion dollars; eight of the twenty-six are oil companies,[2]

In 1900, less than 8 per cent of the power and heat requirements in the United States were furnished by oil and gas; by 1918, 14 per cent; by 1940, 44 per cent; and it is now estimated that over 65 per cent of the national supply of energy is furnished by natural gas, gasoline, and other petroleum products.

Although experiment and application proved before the turn of the century that oil residues could profitably replace coal in the fireboxes under boilers, only 4 per cent of the merchant tonnage and navies of the world were powered by oil in 1914. This percentage increased to 19 by 1920, to 75 by 1944, and approximates 90 per cent at the present time.

The automobile furnished the chief impetus for the rapid expansion of the petroleum industry. The number of passenger cars in the United States increased from 23,000 in 1902 to 902,-000 in 1912, and the number of trucks from none to 41,000 in the same period. Gasoline sales mounted from 5,787,000 to 20,300,000 barrels during that time, and in 1915 gasoline replaced kerosene in consumption of oil-refined products. Passenger cars on the road numbered 6,751,000 in 1919; trucks, 974,000. In 1930, registrations were 23,164,000 and 4,653,000, respectively; in 1940, they were 27,519,000 and 6,849,000. In 1935, the American Petroleum Institute, as a result of a reliable survey, estimated there would be 35,000,000 passenger cars on the roads of America in 1960; but by 1949, that number was exceeded by 1,000,000 cars.[3]

Marquis James has described the evolutionary growth of the retail marketing of automobile fuel, of which there were approximately 425,000 outlets in the United States by 1940:

[2] "The Fortune Directory of the 500 Largest U. S. Industrial Corporations," *Fortune* (a special insert), Vol. LIV, No. 1 (July, 1956); also, "Biggest Year for Big Business," *ibid.*, 88.

[3] "A Survey of Present Position of the Petroleum Industry and Its Outlook toward the Future," American Petroleum Institute, 27.

The filling station and the filling-station pump, the badges of an oil company destined to become most familiar to the average American, came into being by a process of quick evolution. At first the motorist filled his machine by means of a can and a funnel. Most of the gasoline was so likely to be contaminated with dirt that it worked best if strained through a piece of chamois. The quality of gasoline was improved, and also the method of getting it into an automobile. Soon, a motorist could drive into a garage, where an attendant filled his tank from a barrel with a handpump and hose. Then came the underground storage tank, and, about 1910, the pump at the curbstone. These curbstone pumps are recognized as the forerunners of the service station pump of today.[4]

In 1914, only one-sixth of a barrel of crude oil was made into gasoline; new and revolutionary refining processes converted 35 per cent to gasoline by 1926; now, more than 45 per cent can be converted. Before the development of the automobile, gasoline was virtually a waste product, although minor quantities had been marketed as "stove naphtha," which was dangerous and too explosive for general use. Refiners developed gear oils and other automobile oils and greases to meet the requirements of automobile mechanisms. Self-starters began replacing handcranking in 1909, and cars came into general year-round use. When the state and federal governments united efforts on roadbuilding programs, the gasoline-powered motor revolutionized transportation practices, and the refining industry responded to the changing needs. The following table shows the shift from the use of kerosene to gasoline:[5]

[4] Marquis James, The Texaco Story, 28.

[5] An adaptation from Table 9, p. 441, in A History of the Petroleum Administrator for War, 1941–45, prepared under the direction and editorship of John W. Frey and H. Chandler Ide.

GASOLINE AND FUEL OILS SUPPLANT KEROSENE
(1000's OF BARRELS DAILY)

Year	Crude Run To Stills (Barrels)	Gasoline & Naphtha (Barrels)	(%)	Kerosene (Barrels)	(%)	Fuel Oil (Barrels)	(%)
1899	142	18	12.7	82	57.7	22	15.5
1904	184	19	10.3	88	47.8	34	18.5
1914	507	79	15.6	126	24.9	244	48.1
1918	893	226	25.3	119	13.3	477	53.5
1926	2,135	745	34.9	169	7.9	999	46.8
1939	3,391	1,526	45.0	187	5.5	1,282	37.8

Not only is the domestic and industrial use of petroleum products an index of the standard of living, but its importance in global warfare was twice proved in the first half-century. Lord Curzon, British secretary of state for foreign affairs, speaking on November 21, 1918, stated: "The Allies floated to victory on a sea of oil"; in World War II, approximately two-thirds of the cargo shipped from the United States consisted of oil products.[6] In World War I, the Allies used 39,000 barrels of gasoline daily; in World War II, 800,000 barrels.

Equally conspicuous has been the rise of integrated companies in the United States (many operate in other countries) —companies that own production; transportation facilities such as pipe lines, barges, and tankers; refineries; and distributing or marketing outlets for brand-name products.

In 1900, the Standard Oil Company and affiliates had only minor competition in the transportation, refining, and marketing of oil products. According to a survey made by a federal agency, the Bureau of Corporations, in 1904 Standard ran to stills 55,-

[6] Wirt Franklin, in testimony offered on March 21, 1946, on "The Independent Petroleum Company," stated: "Sixty-seven per cent of our overseas shipment during the war consisted of oil products to supply our forces and the Allies." See Hearings before a Special Committee Investigating Petroleum Resources, 79 Cong., *Senate Resolution 36* (1946), 158.

698,000 barrels of oil, or 84.2 per cent; other companies ran 10,-470,600 barrels, or 15.8 per cent. Illuminating oil was still the principal refined product. Standard Oil produced 86.5 per cent of it; other companies, 13.5 per cent. Forty years later, twenty-five times as much oil was refined. The total volume of runs to stills was 1,665,684,000 barrels. Twenty-one companies refined 83.51 per cent of this amount. Eight of these had been affiliated with the Standard Trust Company, which was dissolved by a Supreme Court decree in 1911; they accounted for 649,997,000 barrels, or 39.02 per cent. The remaining thirteen, which became major companies after World War I, ran to stills 741,031,000 barrels, or 44.49 per cent. Two hundred and thirty-four independent companies accounted for 16.49 per cent of the business, or 274,656,000 barrels.[7]

The Hepburn Act, national legislation passed in 1906, and various acts passed by state legislatures were early attempts to break the monopoly which Standard and its affiliates held on pipe-line carriers of crude oil. Not until the Supreme Court ruled on the pipe-line cases in 1916 was effective action taken, however, to assure ratable takings from wells in new fields and enforce common-carrier provisions of pipe-line legislation.

The Interstate Commerce Commission, which has jurisdiction over pipe lines engaged in interstate commerce, reported that in 1950 there were seventy-six companies with 80,996 miles of trunk lines and 47,593 miles of gathering lines. These represented an investment of $1,655,972,000. In 1955, the number of companies had increased to eighty-four, operating 89,729 miles of trunk lines and 50,645 miles of gathering lines. The total investment was $2,585,565,000.[8]

The increasing demand for petroleum and petroleum products during the first half of the twentieth century led to more

[7] Testimony of Fayette B. Dow, National Petroleum Association, *in re ibid.*, 187.

[8] Oklahoma City *Daily Oklahoman,* August 5, 1956, Section D, 6.

scientific and deeper testing for sources of supply. Cumulative production from 1859 to 1901 had been less than 1,000,000,000 barrels; by mid-century, more was being produced in any five months' period. Proved reserves at the beginning of the century were 2,500,000,000 barrels. In 1918, when average daily production was 975,000 barrels, proved reserves were estimated at 6,200,000 barrels; in 1935, daily production was 2,730,000 barrels, and proved reserves, 17,000,000,000 barrels; in 1949, accumulated discoveries since 1859 amounted to 63,500,000,000 barrels, divided into 38,900,000,000 barrels of oil produced and 24,600,000,000 barrels of proved reserves. Daily production had risen to 5,500,000 barrels, an annual production of over 2,200,-000,000 barrels.

Few wells at the turn of the century were drilled to a depth beyond 2,500 feet; by mid-century, exploratory drilling had extended below 20,000 feet, and production was not uncommon at depths between 10,000 and 15,000 feet.

D. Harold Byrd, president of the Independent Petroleum Association of Texas, warned a Senate hearing committee on March 19, 1946:

> We are dependent on the major oil companies for our subsistence. They own or control most of the U. S. production, nearly all imports. They own most of our pipe lines and transportation facilities, including oil tankers. They own 88 per cent of our oil refining capacity and over 80 per cent of our crude oil stocks and manufactured products, including gasoline, kerosene and heating oils.[9]

Oil explorations and discoveries in the Gulf Coast–Southwest region and Mid-Continent area have dominated the petroleum industry during the twentieth century; since 1904, more than 65 per cent of the nation's oil supply has come from Texas,

[9] See the "Independent Petroleum Company," *loc. cit.*, 141.

Oklahoma, Louisiana, Arkansas, Kansas, and New Mexico. The shift of production to this area made possible successful competition; at the beginning of the century, the Standard Oil Company controlled more than 80 per cent of the nation's oil industry; by mid-century, major companies—other than former Standard affiliates—and independents were in the majority.

The search for oil attracted prospectors from Pennsylvania, Ohio, and West Virginia to Kansas, Indian Territory, and other states of the Southwest. Chief among these were John H. Galey and James M. Guffey of Pennsylvania, well-known operators and wildcatters, who helped develop the world-famous McDonald field in western Pennsylvania in 1891 and numbered among their holdings there the "Mathew well," with a record production of 850 barrels an hour when it was almost sixty days old.[10]

Galey and Guffey began operations near Neodesha, Kansas, in 1893 and at the same time began pioneer operations and found limited production in Indian Territory at Bartlesville, Chelsea, Muskogee, and Red Fork. The following year they contracted to drill five wells at Corsicana, Texas. Meantime, a shallow field developed near Neodesha, and they sold their producing properties in 1895—some forty wells—to the Forest Oil Company, a Standard affiliate. Two years later the company built a 500-barrel refinery, the first in the Mid-Continent region, at Neodesha. In 1904, Standard began construction of the first pipe line to tap the crude oil of the Mid-Continent area; it extended from Kansas to its refinery at Whiting, Indiana.

The Corsicana pool, opened by Galey and Guffey, had fifty producing wells by 1897, extending through the city limits eastward in an area five miles by two miles, with average daily production of 65,000 barrels. J. S. Cullinan came to Corsicana from Pennsylvania, liked what he saw, interested outside financial backing, and early in 1898 put into operation a 1,000-barrel refinery. Cullinan used oil for surfacing unpaved streets in Corsi-

10 McLaurin, *op. cit.*, 248.

cana; he and a brother, Dr. M. P. Cullinan, successfully experimented with an oil-burning mechanism for locomotives. At Corsicana, too, a practical rotary drill was perfected which was to make possible deeper drilling than could be attained with cable tools. The bit, fishtail in shape, was attached to a revolving drill pipe. At the upper end of the drill pipe was a hollow square section, the "kelley," held and revolved by the rotary table on the floor of the derrick. Technological advances produced the two-cone rotary bit in 1909; the cones, perfected with more durable metals, made possible the drilling of deeper holes in search of petroleum than the use of cable tools alone.

The rotary bit proved itself at Spindletop, where the discovery well erupted with a roar on January 10, 1901, and for ten days, before the well could be capped, flowed an estimated 70,000 to 110,000 barrels of oil a day. A. F. Lucas, a mining engineer, had interested Galey and Guffey in backing his drilling attempt on Spindletop Hill—so called because it is a gentle rise no more than ten feet above the surrounding coastal plain— a few miles south of Beaumont, Texas.

Although more oil was produced than current demand warranted—for a time a barrel of oil at Spindletop sold for three cents, a glass of water for five—the rich discovery, coterminous with the rapid development of the motor age, prompted new uses of the abundant supply. There was one oil-burning locomotive in 1901; by 1905, there were 227, and by mid-century, over 5,500 oil-burning steam locomotives and 11,000 diesel electric units; railroads alone were using 36,000,000 barrels of diesel oil, and ocean-going vessels, 12,000,000 barrels. In 1901, sugar refineries along the Mississippi River began to substitute oil for coal in their fireboxes. By 1950 more than 500,000,000 barrels of residual fuel oils were consumed by industry in the United States. Many of the oil fields opened in the Gulf Coast–Southwest and California produced asphalt and crude oil with an asphalt base; by mid-century the United States was consuming over 5,250,000,000 tons of petroleum asphaltic products annually. The production of carbon black, a by-product of natural gas used primarily by the rubber industry, amounted to 870,000 pounds in 1948. Daily use of liquefied petroleum gases was 2,800,000,000 gallons, and the total demand for petroleum fuels was 2,250,000 barrels daily. Tankers were carrying over 500,000,000 barrels of petroleum and petroleum products from the Gulf Coast to Atlantic ports annually; the number of tankers had increased from 100 in 1900 to 1,955 a half-century later.

In 1901, when total crude oil production in the United States amounted to 69,300,000 barrels, there were less than 18,000 miles of trunk and gathering pipe lines; in 1910, because of the development of productive oil fields in the Gulf Coast and Mid-Continent area, mileage had been expanded to over 40,000. In 1901 there were less than 1,000 miles of pipe lines carrying petroleum-refined products; by mid-century, more than 20,000 miles were used for this purpose.

The search for oil, spurred by the Spindletop discovery, continued unabated; it reached into Wyoming, where the extent of the Salt Creek discovery in the 1890's was not fully exploited because of market limitations until the Elk Basin and Grass Creek fields were opened twenty years later. Producing properties were developed by means of deeper testing, in the Rocky Mountain states near the close of mid-century; by 1955, lands of the most wretched Navajo tribe in the Rocky Mountain basin, west of the Sangro de Cristo Range where the states of Colorado, Utah, Arizona, and New Mexico meet, were the scenes of great drilling activity.

California, already supplying its own marketing needs in the 1880's, during the next fifty years opened one great pool after another: Barksdale field (1894), Torrey Canyon (1896), San Joaquín field (1908), Shells Canyon (1911), and South Mountain (1916). These provided a superabundance of production for West Coast needs and the export trade. Directional drilling developed at Huntington Beach in the 1920's, to produce offshore oil from onshore drilling sites. As late as 1929, some of the pools of the state were drilled with such intensity that the legs of derricks interlocked.

Out of the Mid-Continent area and the Gulf Coast–Southwest, however, came the supply that fed the growing industrial needs of the nation. In the five-year period following Spindletop, flush production was found at Jennings, Anse–La Butte, and Caddo in Louisiana; Powell, Sour Lake, and Petrolia in Texas; and Red Fork and Glenn Pool in Indian Territory. During the ten-year period preceding entrance into World War I, approximately 80 per cent of the oil produced in the United States came from the Southwestern states. During this period, the Cushing and Healdton fields dominated production in Oklahoma; development spread into West Texas and North Central Texas where the K-M-A, Electra, Burkburnett, Hull, Ranger, and Desdemona fields were principal producing areas. In Louisiana, the Vinton, Pine Prairie, De Soto, Crichton, and Pine Island fields were opened.

Activity in the Mid-Continent area and the Gulf Coast–Southwest has continued since World War I, and this region, with 74.4 per cent of the proved reserves of the country, had accounted for approximately 70 per cent of all oil produced in the United States from 1859 to 1955.[11] During the twenties, production from most of the fields here or elsewhere was limited only by the number of wells and the ability of reservoirs to give up oil, since well-spacing was governed only by the amount of money an operator had or could borrow. But it was in this region that proration was established, ratable taking made mandatory,

[11] Paul D. Torrey, "Evaluation of United States Oil Resources as of January 1, 1956," in *The Oil and Gas Compact Bulletin,* Vol. XV, No. 1 (June, 1956), 19–21. Of the proved resources of 30,000,000,000 barrels in the United States, 14,900,000,000 barrels are in Texas; 3,300,000,000 are in Louisiana; 2,000,-000,000 are in Oklahoma; 998,000,000 are in Kansas; 820,000,000 are in New Mexico; and 330,000,000 are in Arkansas.

and compulsory unitization of pools accomplished in the 1930's and 1940's. The Mexia, Westbrook, Panhandle, and Pierce Junction fields were developed in Texas during the early 1920's; these were followed by Big Lake, Nash, McCamey, Yates, Van, Conroe, and then East Texas, the greatest of all, in 1930. Oklahoma development centered at Burbank, Bald Hill, Tonkawa, Cromwell, the Greater Seminole area, and Oklahoma City.

Kansas developed the Gorham, Ritz-Canton, and Vashell fields, and the Hugaton gas field, which spread into the panhandles of Oklahoma and Texas. Louisiana brought in Haynesville, Bellevue, Cotton Valley, Lockport, and the great Rodessa pool near the borders of Arkansas and Texas. Arkansas experienced typical boom-town activity when the El Dorado discovery in 1921 was followed by the opening of the Smackover pool in 1923.[12] The Hobbs pool in New Mexico, developed in 1929 under practices for conservation of reservoir energy recommended by petroleum engineers, is still a principal producer in that state, where activity has recently increased in the Farmington area. Hundreds of other important pools have been developed in the Southwestern states, and drilling activity has increased in the offshore waters of Texas and Louisiana since World War II.

Illinois, where oil was first discovered in 1889, ranked third among the oil-producing states of the nation in 1940, because of its failure to enact conservation legislation enforced by the other principal producing states.

The fight against monopolistic practices, waged with little success in the oil fields of Pennsylvania and courts of the land up to 1900, followed the oil industry into the Mid-Continent and Gulf Coast–Southwest at the turn of the century. Out of Spindletop and its flush production there developed independ-

---

[12] Excellent summaries of the development of the oil industry in the Southwest appear in Gerald Forbes', *Flush Production: The Epic of Oil in the Gulf-Southwest;* and Carl Coke Rister's, *Oil! Titan of the Southwest.*

ent companies with integrated operations that grew and expanded and competed successfully with one another in supplying the nation's needs for petroleum and petroleum products. Early resentment against the Standard Oil Company and its affiliates was reflected in Kansas in 1905, because the company paid premium prices for oil from Indian Territory and constructed a pipe line from that region to its refinery near Kansas City, instead of depending upon Kansas oil to supply its needs. The Kansas Legislature enacted a law that year, which the Supreme Court later nullified, for the construction of a 1,000-barrel state-owned refinery; additional legislation forbade discrimination in the price of oil and rebates, placed a maximum freight rate on oil, and made pipe lines common carriers. In Oklahoma, common-carrier legislation was enacted, and an abortive attempt was made to prohibit the piping of natural gas from the confines of the state. In Texas, the Waters-Pierce Oil Company, a marketing outlet for Standard, was ousted from the state in 1907 on a charge of violating the state antitrust law, and four years later the parent company was driven from the state on a similar charge. In the Mid-Continent region was won the fight to make pipe lines common carriers, and the right of state regulatory bodies to establish ratable takings from oil pools as a means of regulating the production of crude oil was established.

Michigan, where the first oil discovery was made in 1872, ranked seventh in production by 1937. Production began in Montana in 1920, in Mississippi in 1926, and oil has recently been found in Alabama, the Dakotas, Nebraska, and Missouri. Thirty of the forty-nine states (including Alaska) now produce oil. The ten ranking states in the cumulative production of oil from 1859 to 1956 are Texas, California, Oklahoma, Louisiana, Kansas, Illinois, Wyoming, Pennsylvania, New Mexico, and Arkansas.

The remarkable ability of the oil industry to respond to national needs is apparent by examining the demands made upon it in the second world war. During World War I, 1914–18, as demands for motor fuel increased, and with unlimited production, the United States produced only 90,000,000 more barrels of crude oil in the last year of the war than in 1914; in World War II, 450,000,000 more barrels were produced in 1945 than in 1939. In 1918, the United States lacked that reserve energy, that cushion of potential ability to produce, which was built up in the intervening years before World War II through the cooperation of the oil industry and the state regulatory agencies which administered conservation laws. In 1944, we were able to deliver, because of maintained reservoir pressures, about 85 barrels for each 1,000 barrels of reserves. In 1918, with wells flowing wide open, we could make available only 57 barrels out of each 1,000 barrels of reserves.[13]

The *National Petroleum News* on May 22, 1918, included a discussion on a forthcoming oil shortage created by increased military demands, and noted the possible need of the federal government's creating vast storage depots of gasoline and lubricating oils "against the oil industry not being able to find another Cushing field in the next year or two." However, the great Ranger field in North Central Texas was soon producing 100,000 barrels daily, and the threatened shortage was averted. However, Mark L. Requa, oil administrator under the United States Fuel Administration established in 1917, issued "gasless Sunday" requests for the region east of the Mississippi River on August 27, 1918. Even this restriction was removed after the Armistice, and the Fuel Administration ceased operations on June 30, 1919, but not before Josephus Daniels, secretary of the navy, made a plea for "the nationalization of oil for the future protection of American interests on the sea," at the annual dinner of the American Society of Naval Engineers.[14]

[13] Testimony of J. C. Hunter on "Investigation of Petroleum Resources," June 19, 1945, *Senate Resolution 36, loc. cit.*, 94.

[14] *National Petroleum News*, 6; and June 4, 1919, 19.

In lieu of nationalization, however, the federal Congress enacted a tax policy on depletion to assure extension of exploratory efforts. State conservation laws were passed, and the Interstate Oil Compact Commission was formed. Petroleum engineers inaugurated practices to conserve energy and eliminate waste, the Bureau of Mines established the practice of making monthly forecasts of market demand, and Congress enacted a law to control the transportation in interstate commerce of illegally produced oil. The United States increased the annual production of crude oil by 294 per cent between 1918 and 1941, from 355,-928,000 barrels to 1,402,228,000 barrels. Oil reserves were esti-mated in 1918 to amount to 6,200,000,000 barrels, compared to 20,000,000,000 barrels in 1944; natural gas reserves were estimated to be 13 trillion cubic feet during the last year of World War I, and 130 trillion cubic feet by 1944. During the same period, refining capacity increased from 1,186,000 barrels to over 5,000,000 barrels daily, the gasoline yield per barrel increased from 25 per cent to 44 per cent, and pipe lines were lengthened from 50,000 miles to 140,000 miles.[15]

Before the United States entered World War II, steps were taken to mobilize the oil industry to meet pressing and future needs engendered by the conflict. On May 28, 1941, the President authorized the creation of the Office of Petroleum Co-ordinator for National Defense under Harold L. Ickes, secretary of the interior; on April 20, the agency was changed to the Office of Petroleum Co-ordinator of War; and after December 2, 1942, it was designated the Petroleum Administration for War.[16]

The dynamic, outspoken Secretary, a controversial national figure from the day the New Deal began, later described the voluntary committee arrangements made with members of the industry:

[15] See "A Preliminary Report of Its National Oil Policy Committee to the Petroleum Industry War Council," in *Senate Resolution 36, loc. cit.,* 71–72.

[16] Northcutt Ely, "The Government in the Exercise of the War Power," in *Conservation of Oil and Gas,* edited by Blakely M. Murphy, 664–80.

I want to make one thing emphatically clear. These are *not* rubber stamp committees. Anyone who knows oil men will understand that "yes-ing" is contrary to their nature. And anyone who knows me is aware that yes-men are among my heartiest aversions. Let us be frank. Our early meetings with these committees were not love feasts. There was, as I have acknowledged freely, an atmosphere of suspicion. The oil men were not quite sure what I was up to. They suspected it was something of which they wouldn't approve. They were not quite certain that there wasn't an entrapment somewhere. They were more than a little skeptical as to the workability of the committee plan. The sum total of these factors was an undercurrent of resistance. It distinctly was not a case of love at first sight. Nor at second sight, either. So, during the "honeymoon," there was a good deal of sparring—and some slugging. And then, gradually, there began to dawn the realization that this wasn't purely another marriage of convenience. Maybe this fellow, Ickes, wasn't dealing from a cold deck, after all. And maybe these oil magnates knew how to work in harness. There was a general awakening.[17]

Ickes appointed seventy-two leaders representing all branches of the industry to the Petroleum Industry War Council. The country was also divided into five regions: East Coast, Middle West, Gulf Coast–Southwest, Rocky Mountain, and Pacific Coast. Industry committees were set up in each region to consult with and advise the Petroleum Administration on principal oil operations—production problems, natural gas and natural gasoline, refining, transportation, and marketing. The chairmen of these committees together with a general chairman of the region constituted a general committee to work upon problems that involved more than one of the enumerated functions and to co-ordinate industry activities within the region. These committees, in turn, worked closely with the Petroleum Industry War Council.

17 Harold L. Ickes, *Fightin' Oil*, 76.

The agency in 1942 also formed the National Conference of Petroleum Regulatory Authorities, composed of representatives from each of the oil-producing states, who met regularly to consider and propose the solution of problems related to the war effort.

The committees faced a tremendous task. The United States and Canada, with less than 7 per cent of the world's population, used 62 per cent of its oil; per capita use in the two countries was 353 gallons annually while per capita use in Russia was 41 gallons, in Asia, 4 gallons, and in Europe, 31 gallons annually. Private automobiles had increased in the United States from 6,000,000 in 1918 to 27,500,000 in 1940, and commercial vehicles from less than 1,000,000 to approximately 7,000,000. Seventy-one and two-tenths per cent of the automobiles of the world were registered in the United States. It was necessary for the committees to plan effectively to maintain domestic transportation needs for defense and war activity as well as to provide for the military.[18]

The need for petroleum and petroleum by-products for war purposes had increased tremendously. The United States Navy had only two battleships that used fuel oil exclusively when World War I began; by the early 1930's, the Navy was entirely oil-burning, and during a period of seven weeks in June and July, 1944, the United States Fifth Fleet alone used 630,000,000 gallons of fuel oil. The assault of the *Luftwaffe* upon Britain, which began in September, 1940, emphasized the importance of air power, of lubricating oils, and of aviation gasoline to keep fighter and bomber planes aloft. During the last six months of World War I, the entire Allied Air Force burned less than 12,-500,000 gallons of aviation gas; during one month of attacks on Japanese shipping and installations in 1944, the Far Eastern

[18] Ralph K. Davies, on "Wartime Petroleum Policy under the Petroleum Administrator for War," *Senate Resolution 36, loc. cit.,* 232. Davies, vice-president of Standard Oil of California, was deputy administrator to Ickes.

Air Force burned 143,257,000 gallons; and when the Ninth Air Force was bombing Germany daily, an average of 634,000 gallons of 100-octane gasoline was used on each foray. The drive of Panzer divisions across the Low Countries and France in May and June, 1940, stressed the importance of motorized units; an armored battalion required 17,000 gallons of gasoline to move one hundred miles. In order to meet land and air needs, United States refineries by 1945 were supplying military forces with 35,000,000 gallons of all types of gasoline every day. In World War I, military forces of the Allies and the United States used less than 39,000 barrels of gasoline daily; in World War II, more than 800,000 barrels were used daily.[19] To meet this enormous demand, refineries increased the throughput of crude oil from 3,800,000 barrels daily in 1941 to over 5,000,000 barrels daily in 1945 and boosted the production of 100-octane aviation gasoline from 40,000 barrels to 514,000 barrels daily.

Before the war, few people realized how far refining interests had gone beyond the mere splitting of a barrel of oil into the fractions native to it, or how fully oil and gas processing had become a chemical industry. Under special war powers granted Petroleum Administrator for War Ickes and Deputy Administrator Ralph K. Davies, refining companies manufactured what the government ordered—products, in many instances, peculiar to the war emergency. In addition to basic products, such as regular and aviation gasoline, kerosene, butane, propane, gas oils, fuel oils, lubricants, waxes, and asphalts, the war brought great demands for toluene, butadiene feed stock, isobutene, butylene, and other new and important chemical by-products.

In order to meet military and civilian demands, the petroleum industry produced 5,800,000,000 barrels of crude oil from January, 1942, to August, 1945, and over 13 trillion cubic feet of natural gas, equivalent in heat value to 2,000,000,000 barrels

---

[19] "Report on Petroleum Economics by the Petroleum Industry War Council," *Senate Resolution 36, loc. cit.,* 13–15.

of oil. Natural gasoline production was increased from 150,000 barrels to 315,000 barrels daily, and liquefied petroleum gas, sold before the war primarily as bottled gas to domestic users in areas where other gaseous fuels were not available, was used in making high-octane gasoline, synthetic rubber, and chemicals. Production was increased from 462,852,000 gallons in 1941 to 4,832,722,800 gallons in 1945.[20]

Six months before the United States was drawn into the conflict, Ickes warned domestic users, particularly home-owners in the Northeast who were converting heating units from coal to oil, of an impending fuel shortage in event of war. The rationing of kerosene, heating oil, and residual fuel oil was put into effect on October 22, 1942; gasoline rationing had already begun along the Atlantic seaboard on May 15, as a means of reducing civilian use of highways; it became nationwide on the first of October, as an indirect means of forcing the conservation of rubber.

The problem of logistics or supply occurred in the early months of the war and became the chief obstacle to be overcome through industry and government co-operation. Prior to 1942, over 92 per cent of the oil delivered to the Atlantic seaboard was transported by tankers along the open sea-lanes from the Gulf Coast–Southwest and the Caribbean area; by 1943, less than 12 per cent of the oil delivered to the eastern seaboard came by tanker. In the spring of 1942, German submarines began sinking tankers within sight of the coast: "wolf-packs" ranged off American waters in search of oil-carrying prey. Official records describe their activities as follows:

The beaches of New Jersey, Delaware, and Florida furnish almost the only evidence of enemy destruction in the continental United States. Visible along the high-water mark for miles on these beaches years after 1942 was a winding black ribbon of oil, and just off-shore lie the hulks which mark the terrible

[20] Frey and Ide, op. cit., 228–43.

ordeal of the first war year. They are grim reminders that the Nazis were in almost undisputed possession of the waters of the Atlantic Coast during the early months of the war.

In the first two months after Pearl Harbor, the submarine caused little damage. Only seven U. S. tankers were sunk, two of these on the West Coast. Tanker delivery continued at a normal rate during December and dropped slightly only in January, 1942. Then, in February, the wolf packs struck in earnest. From the Caribbean to Cape Race, no spot was safe. Twelve U. S. tankers went to the bottom in March, the same number in April, and 14 in May. . . . From then on, until the last year of the war, tanker deliveries were an insignificant factor in supplying oil needs to the East Coast.[21]

Tanker shipments to the East Coast dwindled from over 1,421,000 barrels daily in December, 1941, to 173,000 barrels daily in May; the low point of tanker delivery was reached in April and May, 1943, with an average of 63,000 barrels a day. Railroad facilities, pipe lines, barges, and lake tankers were enlisted in the movement of petroleum products so vital to war needs. The change in transportation methods from the Mid-Continent area and the Gulf Coast–Southwest to the East Coast is revealed by the following table:[22]

PETROLEUM DELIVERIES TO THE EAST COAST
(1,000 BARRELS DAILY)

	1941	1943	1944
Tankers	1,421.0	159.5	275.7
Tank Cars	35.0	852.0	646.1
Pipe Line	54.0	266.9	662.7
Barges and Lake Steamers	28.0	112.0	127.6
	1,538.0	1,390.4	1,712.1

There were approximately 115,000 railroad tank cars in service in 1941, principally engaged in transporting refined

[21] *Ibid.*, 87.   [22] *Ibid.*, 449.

products on short hauls from refineries to bulk stations or retail distributing points. About 19,000 of the cars served the East in "local movements." With the war emergency, entire trains made up of tank cars were pressed into service on "through express" service from the Southwest; by 1943, up to 74,000 tank cars were used each day to move more than 850,000 barrels of crude oil to refinery centers along the Atlantic seaboard.

Railroad transport was also used to move petroleum to the West Coast, although there were few interruptions to tanker service once it was through the Panama Canal. Until May, 1943, railroad tank cars delivered less than 10,000 barrels daily. Although seven railroad lines crossed the Rocky Mountains, all were single track except a portion of one, and all lines were congested with materiel and men bound for the Pacific area. Only 17,400 tank cars could be handled by the railroads, but by June, 1945, they were handling 192,000 barrels of crude oil daily.

In order to accelerate the movement of petroleum to the East Coast, the federal government spent $66,000,000 to provide barges, lake tankers, tugs, and towboats for use on the inland waterways. The Plantation pipe line was completed from Baton Rouge, Louisiana, to Greensboro, North Carolina, and by December, 1942, 158,000 barrels of crude oil were delivered daily. The eight-inch Florida Emergency 198-mile-long pipe line was laid across Florida at a cost of $4,000,000, constructed of used pipe from abandoned fields of the Southwest, to transport 30,000 barrels of gasoline received daily from barges on the intracoastal canal.[23]

In March, 1942, as German submarines were cutting off tanker shipments to the East Coast, Mid-Continent and Gulf Coast–Southwest representatives of the oil industry met in Tulsa, Oklahoma. They recommended that oil deliveries to the East Coast be assured by the construction of a twenty-four-inch pipe

[23] George A. Wilson, "Wartime Petroleum Supply and Transportation," *Senate Resolution 36, loc. cit.*, 205–12.

line from the Southwest.[24] Since this was a period of remarkable achievement, their idea was quickly adopted. In June, the War Production Board allocated steel for the purpose, and Burt Hull, president of the Texas Pipe Line Company, a Texaco subsidiary, was selected to manage the construction. Surveying parties immediately began to stake out the right of way; actual construction of the twenty-four-inch crude-oil carrying line began in August.

The first section—531 miles long—from Longview, Texas, to Norris City, Illinois, was completed in January, 1943. The first oil through the line arrived at Norris City on February 19, and trains of tank cars along the seven-mile long racks were loaded for the shuttle service to the East Coast. The pipe line cut their hauling distance in half and made possible the delivery by tank car of twice the volume of crude oil in the time consumed. By April, 200,000 barrels of crude oil were reaching Norris City daily through the line.

The eastern extension from Norris City was authorized in October, 1942, and construction was begun in December. The twenty-four-inch line was extended 723 miles to Phoenixville, Pennsylvania, whence two twenty-inch lines distributed the oil to refinery points in the Philadelphia–New York area. The first oil through the Big Inch reached Philadelphia in August, 1943. The main line, completed in 350 days, was 1,254 miles long. Distribution lines brought the total length to 1,476 miles. Capacity of the line was 3,836,000 barrels, and 300,000 barrels daily could be delivered to the eastern terminals. Up to September, 1945, 260,750,000 barrels of crude oil reached the East Coast through the Big Inch. The construction cost, borne by the government, was approximately $78,500,000.

Construction of the Little Big Inch began in April, 1943. This twenty-inch line extended from Beaumont, Texas, to Linden, New Jersey—1,475 miles—with more than 200 miles of

[24] *Ibid.*, 205.

distributor pipe at the eastern terminal. The line was planned and used for the transportation of finished petroleum products, ordinary gasoline, 100-octane aviation gasoline, motor fuel, diesel oil, heating oil, and kerosene, in batches ranging from 250,000 to 1,300,000 barrels. The line was completed in December at a cost of $67,000,000, and on January 26, after thorough testing, a slug of about 125,000 barrels of gasoline, followed by heating oil, entered the pipe at Beaumont. It required over 2,000,000 barrels of petroleum products to fill the line. Its peak daily carriage was 239,844 barrels; by September, 1945, it had transported 107,125,000 barrels of refined products; about 75 per cent of this was for military use. In order to maintain capacity operation, products from fifteen refineries in the Houston-Beaumont area were pumped into the Texas terminal, while deliveries were made to sixteen terminals on the East Coast.

With completion of these two lines, the Plantation line, and other lines financed by industry and government, the movement of crude oil and by-products by pipe line to the East Coast amounted to 740,000 barrels daily in the spring of 1945. This represented 40 per cent of the amount received in that region, in contrast to the 4 per cent or 54,000 barrels carried by pipe lines each day in the eastward movement in 1941.

Pipe-line mileage in the United States increased from 127,351 miles in 1941 to more than 142,000 miles in 1945. It was necessary to reverse 3,317 miles of existing lines in order to convert them to feeder lines connecting with the Big Inch or Little Inch; 439 miles of twelve-inch, fourteen-inch, and sixteen-inch natural gas lines were converted to crude oil carriers; also, more than 2,000 miles of new crude gathering lines were laid to add to the supply of oil flowing to the Big Inch.

All ocean-going tankers owned by private concerns were taken over by the government for use under the War Shipping Administration. In 1939, 26 per cent of the world's tanker tonnage of 17,235,000 dead-weight tons flew the American flag;

145

by April, 1945, despite terrific sea losses during the earlier part of the war, world tonnage had increased to 22,780,000 dead-weight tons, and the United States owned 54 per cent.

Refineries, too, met the increased demand from military use of by-products. Daily crude oil runs to stills increased from 3,-861,000 barrels in 1941 to 5,001,000 barrels in 1945; the production of 100-octane aviation gasoline increased from 40,000 barrels daily in 1940 to 514,000 barrels daily in 1945. In 1944, synthetic rubber production amounted to 836,000 tons; pre-war rubber imports had never exceeded 650,000 tons a year. Civilian demands for heating oil and gasoline were controlled under a strict rationing system, and approximately 20 per cent of the 425,000 retail gasoline filling stations were closed during the emergency.

New laws for oil and gas regulation, state and federal court decisions, and co-operation between industry and government in the field of conservation contributed to the growth and stabilization of the oil industry during the first half of the twentieth century. The following chapters relate how the industry adjusted to the changing conditions.

9

ECONOMIC WASTE

O UT OF THE Mid-Continent area was to come the movement
that ultimately led to measures regulating oil production,
after the repetitive wasteful practices common to all early developments. Most of the losses resulted from the lack of proper
storage and pipe-line facilities, excessive drilling in highly productive fields, and the inability of producers to control the flow
of wells when these wells were first brought in under heavy gas
pressure. Techniques for the control of reservoir energy were
unknown. J. C. Hunter of Abilene, Texas, in presenting a statement on behalf of the Interstate Oil Compact states to the Congressional Petroleum Investigating Committee on March 28,
1941, stated this point of view succinctly when he said: "What
is waste today, according to our present standards, was not
necessarily waste years ago, and you can no more blame the
State conservation authorities or the oil industry for not having
applied present methods 15 years ago than you can blame a
physician today for not having cured his patients with sulfa-
nilamide in 1925 when the remedy was unknown."[1]

The United States Supreme Court in 1900 recognized the
right of a state to limit the waste of gas from an oil well and
stated the doctrine of correlative rights of landowners in a com-

[1] Hearings before a Subcommittee of the House Committee on Interstate
and Foreign Commerce, 76 Cong., *House Resolution 290;* 77 Cong., *House Resolutions 15 and 118* (1941), 51.

mon source of supply. But the rule of capture remained paramount for more than thirty years, and the states were slow to exercise police power over wasteful production.[2] The federal government limited its authority to the leasing of public and Indian lands for exploratory purposes, with regulations pertaining to production therefrom.[3] Also, releases relative to conservation practices were published by the Bureau of Mines after 1910.[4]

Although the wasteful practices of oil development from the time of the Drake well could be cited for any state, the development in Oklahoma is typical of the industry through the 1920's. It was in Oklahoma that the oil industry observed practices recommended by the federal government for oil production on restricted lands of Indians, and many of the publications issued by the Bureau of Mines on conservation practices were the result of experiments in this state.

The extension of exploration of the Mid-Continent field southward from Kansas began in the 1890's: Indian Territory

[2] *Ohio Oil Company* v. *Indiana,* 177 U. S. 190 (April, 1900). The term "correlative rights" means that each landowner has obligations to the other landowners not to injure the oil or gas reservoir, or to take an undue proportion of the production.

[3] The Geological Survey in 1900 recommended that Naval Reserves be created and withdrawn from public lands. The Elks Hill Reserve in Kerns County, California, was established in 1912, and in 1915, the Teapot Dome Reserve in Wyoming.

The Department of the Interior granted a blanket lease of the Osage Nation, Indian Territory, in 1896, and several leases among lands of the Five Civilized Tribes, Indian Territory, in the 1890's. The Mineral Leasing Act of February 25, 1920, later amended, and special acts of Congress relative to restricted Indian allottees or tribes granted the federal government authority over the production of oil and gas from public lands and the lands of Indians.

[4] The federal Bureau of Mines began the publication of special studies pertaining to petroleum in 1910; it was the pioneer in advocating uniform casing programs, the cementing of wells, and the mud-fluid process in drilling. The first school of petroleum engineering was established at the University of Pittsburgh in 1912; not until after World War I did other universities establish four-year courses leading to a degree in this field. See H. C. Fowler, "Development in the American Petroleum Industry," Information Circular, United States Department of Interior, Bureau of Mines, 8–13.

produced 1,366,748 barrels of crude oil in 1904; 8,264,000 in 1905; and in 1906, the year the first big (Glenn) pool was opened, one year before statehood, 18,091,000 barrels were produced.

In twenty-two years after statehood—1907 to 1929—Oklahoma ranked first in oil production among the states eleven times, and was second eleven times. The maximum output occurred in 1927—277,775,000 barrels. The record of natural gas production can never be compiled. Figures published are only estimates; the total would amount to astronomical proportions, for Oklahoma, West Texas, and western Kansas are in one of the most prolific areas of the world. In Oklahoma one estimate of gas waste, nearly synonymous with production, was 150,-000,000,000 cubic feet to 1910; in 1910 through 1912, it was 100,-000,000,000 cubic feet annually, equivalent to 5,500,000 tons of coal. Gas wasted in 1913, when an official of the Department of the Interior said that wastage in the Oklahoma fields was greater than in any others in America, could have supplied the fuel needs of 1,000,000 families.[5]

When the Glenn Pool oil field was developed, there was no market for the gas released. Many wells with initial volumes of 1,000,000 to 30,000,000 cubic feet daily were permitted to flow wide open for a period of one to eight weeks. It was estimated that 50,000,000,000 cubic feet of gas were wasted here by 1912; the waste was so rapid that three years after the discovery well was drilled there was insufficient pressure for the operation of the field. This same wanton waste of gas extended into the Cleveland area, where most of the oil wells produced an immense volume of gas which was almost wholly wasted. A Bureau of Mines technician estimated the daily waste in this field in January, 1913, to be 20,000,000 cubic feet. He reported:

[5] Raymond S. Blatchley, "Waste of Oil and Gas in the Mid-Continent Fields," *Technical Paper 45*, Department of Interior, Bureau of Mines, 21.

While driving through the Cleveland field several very rich wells were noticed flowing oil at a remarkable rate. Many of the wells were heard blowing oil and gas into large flow tanks. The gas could be seen shimmering from the top of the flues on the tanks and often it was so rich in heavy vapors it would sink to the ground. The valleys or gullies through some of the richer parts of the field were full of heavy gas, which seemed to hang like a fog over the ground.[6]

In the Preston field near Okmulgee, a gas well was allowed to blow wild during February and the first week in March, 1910, at the rate of 36,000,000 cubic feet a day, in the hope that it would produce oil.[7] Another gasser with an estimated daily production of 40,000,000 cubic feet remained open for ten days during this period. In the Schulter field south of Okmulgee, much gas was wasted from the time it was discovered in 1910, and by 1912 approximately 4,500,000,000 cubic feet of gas were blown into the air. On the Bruner allotment seven miles northeast of Henryetta, one well alone had blown gas for 296 days at an average rate of 6,000,000 cubic feet a day, causing a total loss of approximately 2,000,000,000 cubic feet. In the Bald Hill area, flow tanks were used to catch the oil while a daily waste of 8,000,000 cubic feet of gas was permitted. One well lost 40,-000,000 cubic feet in ten days. Many leaky gas lines caused additional waste. The Bureau of Mines expert noticed an ingenious use of the escaping gas by an operator in this field—

[6] *Ibid.*, 37.

[7] The practice of permitting gas wells to flow uncontrolled in the hope that they would begin spraying oil continued in many states into the 1930's. An investigator for the Bureau of Mines reported in 1929 that in the Cook Pool, Shackelford County, Texas, two wells were drilled that blew dry gas at the rate of 250,000,000 cubic feet a day for thirty-four days before oil was produced at the daily rate of 10 to 26 barrels; that in the Wellington Anticline in Colorado, a well produced 82,000,000 cubic feet of gas daily for thirty-five days before spraying oil; that when the prolific Santa Fé Springs field in California was developed in 1923, 50,000 cubic feet of gas were produced with every barrel of oil. See H. C. Miller, *Functions of Natural Gas in the Production of Oil*, 34–78.

a use, nevertheless, that amounted to willful waste. The driller was using the gas pressure to drive a steam engine!

The Healdton field in southern Oklahoma was one of the greatest shallow gas areas in the state by 1914, with wells producing 30,000,000 to 40,000,000 cubic feet daily at 620 feet. One well tested 50,000,000 cubic feet in this sand, and when it was deepened to 750 feet, two additional sands raised the daily production of the well to 100,000,000 cubic feet. The roar of the escaping gas could be heard miles away. The producing sands in this field were somewhat irregular, and wells varied in depth from 600 to 2,100 feet. One well produced 30,000,000 cubic feet of gas daily through the bradenhead and 25,000,000 cubic feet through the six-inch casing, a total production of 55,000,000 cubic feet daily. One sand had a rock pressure of 425 pounds and the other of 300 pounds.

Much heavy, bluish-colored gas was reported escaping from the wells in this field during the first two months of 1914 and settling among the trees and in low places. The city attorney of Ardmore, Oklahoma, wired the United States attorney that some steps should be taken to stop the daily waste of millions of feet of gas. Wells remained open day after day, and the roar of escaping gas could be heard for miles. Visitors in the Healdton area were warned not to come near the field smoking or driving an automobile, and when one great producer got out of control on January 30, automobiles were flagged down, their motors stopped, and teams towed the cars into the field in order to lessen the probability of igniting the gas.

Another great oil field opened when a well was completed in the Cushing area on March 12, 1912. Other locations were immediately established in the area; and in December, 1913, the prolific Bartlesville sand was tapped and in less than one and one-half years, it yielded 85,000,000 barrels of oil. By July, 1915, 800 wells had been drilled into the Layton and Wheeler sands, and 950 wells into the Bartlesville sand. The

Cushing pool completely dominated the nation's oil industry from June, 1914, until our entrance into the European war; it attained its maximum production of 300,000 barrels of crude oil a day in April, 1915.

The oil in this area was confined under very great gas pressure. A number of the wells produced 60,000,000 cubic feet of gas daily that had a pressure of more than 1,200 pounds to the square inch. One of the wells was so great in volume of gas and pressure that it got completely out of control, and the roar made by the gas as it came from the well could be heard for at least six miles. More than 300,000,000 cubic feet of gas were wasting daily from this field by October, 1913, or more than 100,000,-000,000 cubic feet of this ideal fuel in a year's time. This is equivalent to about 5,500,000 tons of coal, and it would have met the wants of nearly 1,000,000 families for one year.

Little effort was made to curb this enormous waste. The Cushing gas field extended over an area ten miles in length and three miles in width, and was underlaid by eleven producing sands. One well during January and February, 1913, wasted 500,000,000 cubic feet of gas. On April 10 of that year a well in the Wheeler sand was allowed to go wild, and it wasted 37,000,000 cubic feet daily through an eight-inch hole. It was shut on April 17, after wasting 250,000,000 cubic feet. Another gasser wasted 30,000,000 cubic feet daily for fifty days, or about 1,500,000,000 cubic feet. Five wells wasted a total of 126,000,-000 cubic feet during April and May, 1913. One of the first wells drilled in the North Cushing field in 1914 wasted an average of 14,000,000 cubic feet of gas each day for sixty-seven days, or a total of 938,000,000 cubic feet. A little later the same well struck another sand and wasted about 40,000,000 cubic feet of gas each day for seven days. From these two sands this well wasted 1,218,000,000 cubic feet of gas, which is equivalent to about 60,000 tons of coal. This represented, however, a small percentage of the gas wasted in the field. Experts of the Bureau of Mines estimated that gas wasted from four wells in this field in a month's time amounted to 6,000,000,000 cubic feet, the equivalent of 250,000 tons of coal, or enough to supply 50,000 families for one year. Less than 10 per cent of the gas in the Cushing field was utilized, and in 1913, 1914, and the first half of 1915, the waste of gas in this field amounted to 200,000,000,- 000 cubic feet. The Bureau of Mines estimated that the daily waste of gas in Oklahoma by 1915 was equivalent to at least 10,000 tons of coal daily, and it was believed that at least 80 per cent of the loss was preventable. The gas wasted in Oklahoma between 1910 and 1915 was estimated to be worth more than all the oil produced during that period; enough gas was wasted to last Oklahoma thirty years.[8]

While this enormous waste in gas was being condoned by oil-men and an unthinking public alike, there was much waste brought about by the overproduction of oil. Operators in the Glenn Pool, Healdton, and Cushing areas were the chief offenders. Here, flush production spurred on by lessors, lessees, royalty owners, and drillers anxious to protect their rights under the rule of capture flooded and broke the market. Markets were unable to absorb the small percentage of oil moved from the fields, and although pipe-line extensions were quickly made into each developed area, independent operators complained that the carriers discriminated in favor of the major producers. A pipe-line company, also engaged in the business of purchasing crude oil, held a producing lease at Glenn Pool when there was no other marketing outlet. This company produced its well to capacity and carried its oil to market at a time when its neighbors were able to dispose of only a small part of their production. These less fortunate operators stored their production in gullies, creeks, and earthen reservoirs while wooden and steel tanks

---

[8] A. G. Heggem and J. A. Pollard, "Drilling Oil Wells in Oklahoma by the Mud-laden Fluid Method," *Technical Paper 68,* Department of Interior, Bureau of Mines, 14.

were under construction. These were soon full and overflowing with oil. Hundreds of wild ducks and geese, flying from the north to their winter feeding ground in the south, were attracted by the ponds and trapped in the oily lakes. Much of this oil might have remained in the ground until needed for commercial demands and until it could be sold for a price commensurate with its value, but operators were governed by the rule of capture, and each made as big a grab as possible.

Oil leaves a heavy residue which collects when in storage. At this time, the residue was either burned periodically or left— a dirty, sticky mess—to flow down gullies toward the nearest stream. A Bureau of Mines expert found much waste of this nature in the fields of northeastern Oklahoma in 1912. He estimated that there were 12,000 barrels of residue in a 25,000-barrel pond near Cleveland, while southwest of the city he noticed an earthern reservoir containing 25,000 barrels of residue. About 250,000 barrels of oil were stored here because of the activity at Glenn Pool in 1907, but evaporation reduced the oil to a thick residue.[9]

Heavy rains in the Healdton oil field caused several dams of the earthen storage tanks to wash out in May, 1914. Fully 150,000 barrels of oil floated down Bayou Creek, 50,000 barrels soaked into the earth, and an additional 50,000 barrels evaporated during the summer months. At the same time, there was an estimated daily loss of 100,000,000 cubic feet of natural gas. Observers noted that the ground was literally soaked with oil and the air was saturated with natural gas. One person who drove over the field before the rain said his auto was axle-deep in oil at times. A committee of independent producers accompanied Cato Sells, commissioner of Indian affairs, through the Healdton field on May 29, 1914. All portions of the producing field were visited and investigated. They witnessed immense quantities of waste oil burning in Bayou Creek, heavy columns of smoke

[9] Blatchley, *op. cit.*, 23.

marking the creek for miles. At this time, there were 400,000 barrels of oil in storage, 60 per cent in earthen tanks which had been located on higher ground following a disastrous flood earlier in the month. One earthen tank had just been completed that would hold 200,000 barrels. Commissioner Sells was supplied with data which showed that 40,000 barrels of oil were being produced daily, of which the Magnolia Pipe Line Company could carry only 8,000 barrels. Healdton oil operators claimed that the loss from seepage and evaporation amounted to 25 to 50 per cent. Sells was particularly interested in this field because some of the producing wells were on Indian allotments. After visiting the Healdton field, one member of the official party said: "There is more oil flowing daily down the Bayou in this field and Tiger Creek in the Cushing field than is produced in many of the well known fields. The thousands of barrels of oil in Bayou Creek, a total loss, showed more than any other thing the great need for market facilities."

Throughout the summer months of 1914, thousands of barrels of oil continued to flow down Bayou Creek. A six-inch rain in September washed out a number of dams that held big earthen storage, and the oil spread over the lowland farms. A most destructive fire, started by lightning, had broken out in the field during the preceding month, destroying much of the above-ground storage. A 40,000,000-cubic-foot gas well, two flowing oil wells, six 55,000-barrel steel tanks, several wooden ones, and ponds of oil were consumed. In October, heavy rains washed out many earthen ponds spreading their contents over the lowlands and leaving a film of oil to destroy the crops.

A similar condition existed in the Cushing field by 1914. Besides the millions of feet of gas wasting into the air, many more oil wells were producing than were necessary for maximum recovery, and great quantities of oil were piling up in above-ground storage as the export trade became limited by the outbreak of war in Europe. Every new well opened in the highly

155

productive Bartlesville sand added to the mounting surplus. The oil producer was unable to stop drilling activities, however, because he wanted to protect his investment, and he knew that if he did not take the oil from under the ground, his neighbor probably would. Since the rule of capture prohibited underground storage by a conservation-minded producer and there was no outlet because of limited transportation facilities and market demand, he was forced to store his oil. Oklahoma ranked first among the states in the production of crude oil in 1915, producing 25 per cent of the nation's output and 55 per cent of the high-grade oil marketed in the United States. The Cushing area alone supplied 17 per cent of the nation's production and 30 per cent of the high-grade oil, while enormous surpluses of this oil were stored aboveground. Ten million barrels were stored here by September, 1914, and it was estimated that at least 25 per cent of this amount would evaporate within a year's time. Pipe lines capable of moving 40,000 barrels of oil to market daily, when running at full capacity, were operating five months later, but storage, by then, amounted to 20,000,000 barrels. If not another barrel of this high-grade fluid had been turned into the tanks, it would have taken more than a year and a half for the pipe lines to carry the oil aboveground to market. And all during the winter and spring seasons, wells in this area produced more than 250,000 barrels daily. Evaporation, seepage, and runoff took a heavy toll. A Bureau of Mines official estimated that when the Cushing field was producing 174,000 barrels of oil daily early in 1914, some 25,000 barrels were allowed to run into the Cimarron River. The enormous gas pressure often caused wells, when completed, to blow the oil high into the air before they could be controlled. An observer once noted that a cotton field was covered with oil for one-fourth of a mile in all directions.

There are wastes of natural gas and oil other than the obvious types apparent in flush production. The most evident waste of natural gas occurs when a well is allowed to go "wild" after

being drilled into a gas-bearing stratum, but A. G. Heggem and
J. A. Pollard of the Bureau of Mines wrote in 1914:

> Other wastes of perhaps as great importance occur in drill-
> ing by the usual methods. Gas is often allowed to escape freely
> while a well is being drilled through a gas-bearing sand in
> search of oil. In some wells an attempt is made to save the
> gas by shutting it in with a string of casing, having a packer at
> the bottom and a stuffing-box casing-head known as a Braden-
> head, at the top. Between the packer and the casing-head
> there is usually a large amount of open hole; that is, a hole in
> which the gas is confined in direct contact with the strata
> penetrated, and the strata may be porous. Much of the gas
> enters the more permeable strata, sometimes forcing its way
> to great distances and is lost, so that when the well is opened
> at a later day the available supply of gas has decreased to such
> an extent that it is of practically no value to the owner. This
> subterranean movement of natural gas sometimes leads to the
> rejuvenation of exhausted gas sands, which has been observed
> in some of the older fields, and in other instances has consti-
> tuted a formidable danger by establishing a "stray sand," the
> unexpected encountering of which at another well may result
> in a gas fire with attendant loss of life and property.[10]

Experts of the Bureau of Mines in 1916 reported several
instances of underground waste.[11] In the Cushing area, gas of
high pressure reached gas of lower pressure and shallow water-

---

[10] Heggem and Pollard, "Drilling Oil Wells in Oklahoma," *op. cit.,* 12–13.
In recent years, in order to meet industrial and domestic demands for natural gas,
heaviest during the winter months, vast underground storage has been made in
old fields. There were 167 storage pools in 1953, principally in Illinois, Michigan,
Ohio, Pennsylvania, West Virginia, and Oklahoma. It was estimated that 404,-
838,000,000 cubic feet of gas were placed in storage, and 246,802,000,000 cubic
feet were withdrawn, totaling 158,036,000,000 cubic feet of net storage that year.
See *Mineral Yearbook: Fuels,* Department of Interior, II, 317.

[11] William F. McMurray and James O. Lewis, "Underground Wastes in Oil
and Gas Fields and Methods of Prevention," *Technical Paper 130,* Department
of Interior, Bureau of Mines.

bearing sands, and within a few days gas began to erupt from two wells being drilled 500 feet away and from crevices in the bottom of a ravine still farther away. A few months later another well being drilled near by unexpectedly encountered gas in the water-bearing sands, and the rig was burned down. At Cushing, too, the conservation agent of the Corporation Commission found that waste of natural gas from the bradenhead was frequently concealed by burying the pipe running from it; by releasing the gas in sand, rock piles, or under the waters of the Cimarron River; and by connecting with the flow line and allowing the gas to escape with the flowing oil. Sometimes a line from the bradenhead was run into the flow tank to disguise the fact that gas, and not oil, was flowing. In Osage County, a well producing 22,000,000 cubic feet of gas daily was under a pressure of 900 pounds, which caused the gas to cut under the $8\frac{1}{4}$-, 10-, and $12\frac{1}{2}$-inch casings and come up between the $12\frac{1}{2}$- and 16-inch casings to the surface. Cement between the 16-inch casing and the walls of the hole and wooden pegs between the $12\frac{1}{2}$- and 16-inch casings caused the gas to escape from many places within 300 feet of the well, while 600 feet away gas was noticed escaping from the bottom of the creek. Near Loco, in southern Oklahoma, a defective packer caused underground waste that resulted in a drop in the rock pressure of a gasser from 310 pounds to 90 pounds. After the casing was seated, however, the pressure increased to 210 pounds in forty-five days. Near Duncan, two gas wells, 800 feet apart, were drilled into a gas sand. The first well flowed 3,000,000 cubic feet daily; the other 10,-500,000 cubic feet. The second well when deepened 100 feet showed a rock pressure of 310 pounds, but four days later this pressure decreased to 245 pounds while the pressure in the first well increased from 60 pounds to 220 pounds. The gas could plainly be heard feeding into the upper sand.

Another type of underground waste was caused by the premature flooding of the oil or gas stratum by water infiltrating from higher strata through leaky casing, from water cutting under the casing, or from improperly plugged wells which permitted water to leak into the oil sands. There were other types of waste apparent in the producing gas fields. At Cushing, Cleveland, and elsewhere, there was great waste from the production of oil and gas from the same well when the sands were clearly separated by shale or other strata. The burning of torches or flambeaux night and day was a common practice; the irregular spacing of wells also caused much economic waste. Well-spacing is of more vital importance in a gas field than in an oil field because of the greater ease with which gas migrates from one formation to another. A closer spacing than is necessary increases the possibilities of allowing gas to escape into unproductive formations and thus decreases the volume of the ultimate recovery. Then, too, each additional well increases the water hazard. F. P. Fisher of the Wichita Pipe Line Company recommended to the Corporation Commission in 1915 that one oil well per ten acres and one gas well per forty acres should be the best general average for recovery from the underground reservoirs, instead of the wasteful crowding of wells in an unrestrained desire to get as much oil and gas aboveground as quickly as possible.

Because of market limitations, the owners of wells were at the mercy of the pipe lines even after the state passed laws bearing upon conservation. Near Okmulgee, an operator, while drilling for oil, struck a 40,000,000-cubic-foot gas well. He offered to sell the gas to the gas companies or pipe lines at their own price; they refused to buy it but complained to the county attorney, who forced the owner of the well to cap it to prevent the gas from escaping. Then the pipe-line company proceeded to tap the gas through another well in the same field. In 1912, a like condition had existed in the Collinsville area, where a number of wells were closed when the pressure fell from 540 pounds to 360 pounds. The closed wells, nevertheless, lost much gas from excessive draining, to a company that entered the field and

159

drew gas only from wells near its trunk line. Although the gas wells near the pipe line sold at a ridiculously low figure, the owners of wells farther away were compelled to sell their gas at an even lower figure or lose it entirely.

Many companies, too, furnished gas to industrial and domestic consumers in the growing towns at a flat rate. This was conducive to much waste in consumption. Boom-towns offered gas free or at a very low rate to manufacturers, as an inducement to locate in the community. Typical of such inducements is the following news item taken from the front page of the Nowata *Advertiser* of February 28, 1908:

> After three years of talk the 12 inch pipe line from the Hogshooter gas field to the west is now an actual reality and Nowata is in a position, right now, today, this very minute to supply natural gas for factories at the rate of two cents per thousand feet—the only town in the world which can make such an offer and deliver the goods.
>
> One hundred million feet of gas per day, that is what Nowata has to offer one and all at 2 cents per thousand cubic feet. There are no other fingers in the pie, either. The gas is for Nowata alone. The Kansas Natural has no grappling hooks around the territory.[12]

Four years later, however, the citizens of Nowata found they had insufficient gas for domestic use!

When flush production at Glenn Pool necessitated much aboveground storage, the Interior Department required that earthen storage reservoirs on Indian lands should have dikes fifteen feet high. This was a precaution against the surface waste of produced oil, and the order was issued not with conservation in mind, but to assure the wards of the government that their royalty rights would be protected. Wasteful methods of drilling that permitted the unrestrained flow of gas from sands while

[12] Nowata (Oklahoma) *Advertiser,* February 28, 1908.

penetrating deeper for oil attracted the attention of the Interior Department, and in the summer of 1913, the Bureau of Mines demonstrated that wells could be drilled and gas formations sealed so as to prevent the further waste and escape of gas.[13] It was recommended that each gas-bearing stratum be sealed as it was encountered, by drilling with the hole full of mud-laden fluid. This process was introduced into the Cushing field at the request of the Interior Department, but most of the operators remained indifferent to it.

The Oklahoma Independent Oil and Gas Association, organized in February, 1913, succeeded in securing passage by the state legislature of *House Bill 723*, which made gas pipe-line companies common carriers and common purchasers.[14] Another act of the 1913 legislature provided for ratable production from common reservoirs of natural gas; correlative rights of surface owners in the common reservoir were recognized, and each producing well was limited to 2 per cent of its daily natural flow unless good cause could be shown to the Corporation Commission, the state regulatory agency, for making a greater withdrawal.

Officials of the Interior Department and the Corporation Commission of Oklahoma began to work in closer co-operation after the flush production in the Cushing and Healdton fields

---

[13] See Heggem and Pollard, "Drilling Oil Wells in Oklahoma," *op. cit.*, and "Mud-laden Fluid Applied to Well-Drilling," *Technical Paper 66,* Department of Interior, Bureau of Mines.

[14] The first legislature, on December 21, 1907, enacted a law which forbade the transportation of natural gas outside the boundaries of the state. This law, which theoretically was a conservation measure in the sense that gas produced from Oklahoma wells should be consumed or wasted only in Oklahoma, was immediately attacked in federal court and declared unconstitutional. See *Kansas Natural Gas Company* v. *Haskell et al.,* 172 Federal Report 545 (July, 1909) ; and *West* v. *Kansas Natural Gas Company,* 221 U. S. 229 (May, 1911).

An act of the legislature in January, 1909, made all pipe-line carriers of crude oil common purchasers of oil, provided for ratable taking, and made it unlawful for any corporation engaged not as a common purchaser in transporting crude oil to operate any oil wells or leases within the state.

in 1912 and 1913. On August 1, 1914, Congress provided six additional oil and gas inspectors to supervise oil- and gas-mining operations on allotted lands leased by members of the Five Civilized Tribes, from whom restrictions had not been removed. These inspectors visited Oklahoma and reported enormous losses in the production methods being carried out in the flush fields.

Earlier in the year a delegation of independent oil producers from Oklahoma visited Washington and complained that the major pipe-line companies operating in Oklahoma had conspired to create a monopoly in restraint of trade. The Attorney General admitted that pipe-line companies could control the price of oil, but said that he had been unable to discover how they could be reached by law. As a result of the visit of the delegation, however, Senator Thomas P. Gore of Oklahoma introduced a bill in Congress to prohibit pipe-line companies from owning production, while Senator Robert L. Owen introduced a bill which advocated government ownership of the lines.

Overproduction was, of course, the chief reason for disputes between producers and pipe-line companies. Hardly had their facilities been expanded to handle oil from Glenn Pool when the richly productive Bartlesville sand in the Cushing field was tapped. Fifty to one hundred feet thick, its wells averaged more than 1,000 barrels a day and, because of great gas pressure, promised to keep up their natural flow for many months. Oil operators in northeastern Oklahoma agreed that the pipe-line companies should run all the oil from the less prolific Wheeler and Layton sands, but that oil from the Bartlesville sand should be prorated according to the potential capacity of the producing properties.

Only about 22 per cent of the Cushing field output could be taken, and a mounting surplus of the highest-grade oil ever discovered in the history of petroleum was being stored aboveground in wooden or steel tankage or dammed up in creeks or earthen pits at the rate of 3,000,000 barrels a month. Pipe-line companies operating in the Cushing field asked the Corporation Commission to relieve them from operating or taking any part of the oil from wells drilled within the next four months. Members of the Commission hesitated, but they finally acted in order to "prevent waste of a product which the public has need for and must use."

In its order of July 1, 1914, the Commission made this analogy: "If a forest of timber were being ruthlessly and carelessly destroyed by fire or sawed down and permitted to float down rivers, the public would find some way of stopping such destruction."[15] A few months later overproduction in the Healdton pool almost duplicated conditions of the Cushing area, and the Corporation Commission issued an order to limit production there.

Ten million barrels of oil were stored aboveground in the Cushing area by mid-September, and producers again besieged the Commission with requests to do something toward stabilizing the price of oil. Its members decided that common law gave them the right to prevent waste and ordered that their permission must be obtained before any additional wells could be drilled. Pipe-line companies were relieved as common purchasers from their obligation to take oil from wells drilled in violation of the order. The Commission also sought to establish a minimum price of 55 cents a barrel on oil produced within the state.

By its price-fixing order, the Commission obviously was trying to get producers to leave the oil underground until market conditions would justify its withdrawal. The legality of the order was challenged by the Magnolia Pipe Line Company, which notified the Attorney General and the Corporation Commission on September 30, 1914, that it would reduce the price of oil in the Healdton field on October 6 to 40 cents a barrel. The Commission then ordered that no producer should take any oil from

[15] Order No. 829, issued by the Oklahoma Corporation Commission on July 1, 1914, at Oklahoma City.

the oil sands or from storage and sell it for less than 50 cents a barrel. But because oil in earthen tanks was deteriorating so fast, producers in the Healdton field, ignoring the Commission's order, agreed to accept an offer of the pipe-line company to purchase 250,000 barrels of earthen storage oil at 30 cents a barrel. Other producers were willing to sell thousands of barrels of stored oil below the designated price; hence, this early attempt at price-fixing was unsuccessful.

The Commission's attempt to limit economic waste by prohibiting the drilling or shooting of any more oil wells in Oklahoma without its consent did not go unchallenged. The Quaker and Silurian oil companies each filed petitions in federal court to set aside the restrictive order of the Commission, despite the fact that high-gravity crude oil had declined in price from $1.05 a barrel in April, 1915, to 55 cents a barrel in September. The demand of other oil producers that something should be done to restrict output was insistent, but the old "rule of capture" wrecked plans for voluntary action. Certain producers petitioned Governor Lee Cruce to call out the state troops and shut down every well, but the Governor told the oilmen that the aid of the courts must be invoked before martial law would be declared. Then a sweeping shut-down order was petitioned, but the Commission did not feel empowered to exercise such authority. The price of oil, however, was set at 55 cents, and the pipe-line companies were released from taking oil from any new production.

Again producers offered oil to the pipe-line companies below the set market price, and the Silurian and Quaker oil companies attacked the shut-down order affecting the Cushing field. This attack forced the Corporation Commission to extend drilling exemptions to "all wells necessary to give equitable protection to property from drainage." Production reached 245,000 barrels a day by August 2, 1915, and the price of oil was down to a low level of 40 cents a barrel. Remedial measures stronger than voluntary action obviously were necessary if waste was to be prevented.

The Fifth Oklahoma Legislature, which convened in January, 1915, passed comprehensive conservation measures designed to prevent waste of oil and gas. Perhaps the most important factor leading to the legislation was the conviction of oil operators that the industry needed regulation. The Conservation Statutes of 1915 were proposed, written by, and passed through the influence of the oil lobby.

House Bill 168, approved on February 15, 1915, prohibited waste and production of crude oil when its market value fell below the cost of production. It provided for the ratable taking of crude oil from a common source of supply. It authorized the Corporation Commission to issue rules for the enforcement of the act. The definitions in Section 3 of the bill attracted criticism: "The term 'waste' as used herein, in addition to its ordinary meaning shall include economic waste, underground waste, surface waste and waste incident to the production of crude oil or petroleum in excess of transportation or marketing facilities or reasonable market demand."

House Bill 395, for conservation of natural gas, was approved on March 30, 1915. Section 2 provided "that the term waste as used herein in addition to its ordinary meaning shall include escape of natural gas in commercial quantities into the open air, the intentional drowning with water of a gas stratum capable of producing gas in commercial quantities, underground waste, and permitting of any natural gas well to wastefully burn and the wasteful utilization of such gas." Wells capable of producing 2,000,000 cubic feet daily were considered of commercial quantity. The act included provisions that had been enacted by previous legislatures: a common purchaser clause and ratable taking from a common source of supply.

10

EXPERIMENTS

IN CONTROLLED PRODUCTION

Oklahoma's natural gas act of 1913, which provided for ratable taking from a common reservoir, and the conservation acts of 1915 relating to oil and gas production were the first comprehensive legislation enacted by any state on the subject.[1] The Corporation Commission, a regulatory agency created by the state constitution, was given administrative responsibility, and with the backing of members of the Independent Oil and Gas Operators' Association, a general order containing twenty-eight rules was issued by the Commission, to be effective on September 1.[2]

In addition to incorporating provisions of the legislative act, the order included suggestions from the Bureau of Mines. Its provisions included the protection of fresh, underground waters from pollution, the plugging of dry or abandoned holes, the use of a casing head to control gas under high pressure, and the necessity of sealing off any stratum of oil, gas, or water either by the mud-laden fluid process or by casing and packers when drilling through the sand. Separators were to be used when oil and gas were produced from the same stratum, and the mud-laden fluid process was recommended for confining gas unless

[1] Northcutt Ely, compiler, *The Oil and Gas Conservation Statutes.*

[2] Order No. 937, August 16, 1915, in *Eighth and Ninth Annual Reports of the Corporation Commission of the State of Oklahoma, for Years Ending June 30, 1915, and June 30, 1916,* 287–323.

the bradenhead method could be satisfactorily used. In the use of the bradenhead, separate strings of casing were to be run to each sand. Vacuum pumps could be used only by permission of the Commission. Casing could be ordered raised and reset in mud-laden fluid when it was thought advisable to do so in order to avoid existing underground waste, pollution, or infiltration. Operators were ordered to file well logs and report the plugging of wells to the Commission, which also required a monthly gauge of the volume and rock pressure of all wells producing natural gas. No gas well should produce more than 25 per cent of its daily capacity.

Before this general order was issued, the Ardmore Oil Producers' Association had filed a complaint with the Corporation Commission, as provided by the Oil Conservation Act. They charged there was excessive waste in the Healdton field. Overproduction forced the price of the oil produced there from $1.05 a barrel in February, 1914, to 30 cents a barrel nine months later, and producers who stored surplus production in steel tankage paid 25 cents a barrel for insurance.

The complaint in regard to this waste was heard by the Commission on April 23. General attention centered on disposition of the case because it was the first test of the powers conferred on the Commission by the legislative act to limit production in a field to market demand and to prorate production of crude oil on a percentage of potential basis.[3]

The Commission found that the market demand for oil from the Healdton field did not exceed 15,000 barrels of crude oil a day and that transportation facilities did not exceed 25,000 barrels daily, while production ran between 70,000 and 80,000 barrels a day.

Obviously, a great surplus of unmarketable oil was going into storage. Producers estimated that oil stored in the best

[3] W. P. Z. German, "Legal History of Conservation of Oil and Gas in Oklahoma," in *Legal History of Conservation of Oil and Gas*, 129.

aboveground steel storage depreciated in value 10 to 25 per cent in the first year. The Corporation Commission reported:

> If we consider the minimum amount fixed by the witnesses as waste of oil in steel storage to be 10 per cent of the value and 80,000 barrels of oil were stored each day for one year in the Healdton field there would be a loss of eight thousand barrels per day. Would it not be a spectacle for the public to assemble each morning and witness the destruction of eight thousand barrels of a commodity that is absolutely necessary for the prosperity and business of the country—a commodity which once destroyed can never be replaced by any artificial means known to science.[4]

The Commission ordered that earthen storage be discontinued and that wooden or steel storage be permitted for ten days' production from any well in order to provide for unforeseen contingencies, but that the accumulation should be charged against the prorated share of the producer. The potential production of each well was to be ascertained by a conservation officer who would permit the operator to produce his just percentage of the daily market demand. No more oil was to be produced than could be marketed, and no operator could take more oil that his equitable proportion of potential production.

The Corporation Commission on July 16, 1917, issued Order No. 1299 for the conservation of natural gas and oil. Its forty-one rules consolidated the applicable general rules and regulations, and it remained in operation, with minor additions, until the Oklahoma Legislature revised its conservation laws in 1933. Rule 12, general enough to suffice for any details otherwise overlooked, provided that "all operators, contractors, or drillers, pipe line companies, gas distributing companies or individuals, drilling for or producing crude oil or natural gas, or piping oil or

[4] Order No. 920, June 5, 1915, in *Eighth and Ninth Annual Reports of the Corporation Commission of the State of Oklahoma, op. cit.,* 260.

gas for any purpose, shall use every possible precaution in accordance with the most approved methods, to stop and prevent waste of oil and gas, or both, in drilling and producing operations, storage or in piping or distributing, and shall not wastefully utilize oil or gas, or allow same to leak or escape from natural reservoirs, wells, tanks, containers, or pipes."

Meanwhile, many of the problems of waste faced by the oil industry were being attacked by petroleum production engineers. The Bureau of Mines located a Petroleum Experiment Station at Bartlesville, Oklahoma, in 1918, and its technicians continued and extended the work already undertaken by the Bureau.[5] As its investigations in oil-recovery methods, natural gas conservation, and evaporative losses of gasoline and petroleum, and its reports on the operation of its experimental refinery became better known, operators paid more attention to its recommendations.

Bureau of Mines engineers in 1919 advised operators in the Hewitt field of southern Oklahoma how deep they should bore with the rotary drill; a well near Comanche in Stephens County, Oklahoma, that would otherwise have been a dry hole, was deepened on the advice of a Bureau engineer and came in at 450 barrels a day. At Cushing, where operators found their production threatened by underground water, engineers from the Experiment Station supervised the repair of the wells. This remedy increased oil production 4,504 barrels a day and decreased water production 6,892 barrels a day. A study of evaporative

[5] The Bureau of Mines was created in 1910 for the purpose of increasing health, safety, and efficiency in the mining industries. Beginning with the establishment of the Pittsburgh Experiment Station, the bureau now has fourteen Experiment Stations in different parts of the United States. In 1923, the Petroleum Laboratory was transferred from Pittsburgh, Pennsylvania, to Bartlesville, Oklahoma. Field and laboratory studies of production and transportation of oil and gas are conducted by staffs of the petroleum experiment stations at Bartlesville and Laramie, Wyoming, at the helium plant at Amarillo, Texas, and at the offices in San Francisco, California, and Dallas, Texas. The stations at Bartlesville and Laramie also conduct experiments pertaining to petroleum chemistry and refining.

losses in the Mid-Continent field showed that in 1919 more than 122,000,000 gallons of gasoline evaporated while crude oil was stored for an average of five days on the lease. The Experiment Station, too, made recommendations on the importance of well-spacing for maximum recovery. Investigations were publicized which showed operators how open-flow wells permitted the escape of quantities of gas without contributing to the recovery of oil. Experiments in the use of compressed air to force oil from the sand, first attempted in the oil regions of Pennsylvania in the 1860's, and the determination of proper back pressure engaged the attention of petroleum engineers. Production was increased in a well near Bartlesville by increasing the tubing depth of the well. Investigations were also carried out on pumping wells to determine the best rate of pumping, the proper length of stroke, the best type of pump, and the proper pumping depth to obtain the most oil.

It was generally known by 1920 that about 80 per cent of the oil was not being recovered from the sands. Bureau experts demonstrated how the introduction of extraneous gas into an oil sand would increase the production of oil wells. The practicability of this experiment was shown in the Lyons-Quinn field in Okmulgee and Okfuskee counties in Oklahoma in 1923, where gas forced into the Lyons oil sand increased production by about 75 per cent.[6] Gas to the value of $144,400 was used in the operation, but it was instrumental in recovering oil valued at $1,525,-000. When the great Chickasha gas field, discovered in July, 1922, developed wells capable of producing 1,000,000,000 cubic feet of gas daily, Bureau officials made recommendations that were followed in guarding against a loss of gas by migration or by water infiltration.[7] Less than 2 per cent of the potential daily production was taken from the wells.

[6] M. J. Kirwan, "Effects of Extraneous Gas on the Productivity of Oil Wells in the Lyons-Quinn Field of Oklahoma," Department of Interior, Bureau of Mines (June, 1924).

In 1920, Bureau of Mines officials called a meeting of gas producers, attorneys, and others from Oklahoma, Texas, Louisiana, Arkansas, and Pennsylvania to discuss means of conserving the natural gas supply, but the industry in general continued through the following decade with little thought of conservation. By the time the Seminole field in Oklahoma was developed in 1926, however, oil producers had learned the practicability of pumping gas or air into wells in order to draw more heavily upon the oil in the sands. It was estimated that by March, 1927, more than 150,000,000 cubic feet of compressor capacity were installed in that field, and more footage was being built. By the utilization of gas or compressed air lift, it was possible to produce an amount of oil in ten or eleven months that would ordinarily have been spread over four to five years.[8] But here as elsewhere at this date, this use of natural gas was regarded as an operating economy rather than as a conservation measure.

In other fields and in earlier years, the natural gasoline contained in "wet" gas blown from oil wells had been wasted. Early in the century a successful plant was installed in West Virginia to recover gasoline from dry gas, and by 1927 it had become customary to utilize the "wet" gas unavoidably produced from oil wells by processing it through a plant for the extraction of its valuable gasoline content. Residual dry gas was marketed or compressed and reinjected into the reservoir from which it was originally produced, there to aid in maintaining pressure in order to drive more oil through the underground reservoir to the well bore. *The Oil and Gas Journal* of December 1, 1927, estimated that gasoline extracted from the casing-head gas at the wells during the previous year amounted to approximately 1,-600,000,000 gallons.

Consumption of gasoline rapidly increased with the rise of

[7] M. J. Kirwan and T. E. Swigart, "Engineering Report of the Chickasha Gas Field," *ibid.* (May 1, 1923).

[8] C. N. Swarts, C. R. Bapp, W. S. Norris, "Petroleum Engineering Report on the Seminole Pool," *ibid.* (July, 1928).

171

the automobile age; until 1915, more kerosene than gasoline was extracted from crude oil. Increased demand for the latter product prompted improved refining techniques which, in turn, led to the rapid expansion in the 1920's of a peculiarly American institution—the retail filling station. Although there were only 15,000 filling stations in 1920, by the end of the decade over 121,000 were strategically located in cities and along the expanding highway system of the country.[9] In 1912, William M. Burton perfected the Burton process for cracking crude oil molecules to produce gasoline; his process, patented by the Standard Oil Company of Indiana, was the only one in commercial operation by 1914. Cracking converts a less-wanted product into one of higher worth; the molecules of the heavier fractions of crude oil are separated and then reunited into particles composing lighter fractions. The Bureau of Mines in 1915 offered to refiners at no cost a cracking process developed by its technicians, and in 1918 Ralph C. Holmes and Frederick T. Manley of the Texas Company developed a process begun by Joseph H. Adams.[10] Other refiners adopted these processes or improved their own practices, and the amount of gasoline extracted from one barrel of average crude oil containing 42 gallons increased from 6.5 gallons in 1914 to 18.4 gallons by 1929.[11] By mid-

[9] "Investigations of the Concentration of Economic Power," *The Petroleum Industry*, Hearings before the Temporary National Economic Committee, 76 Cong., *Public Resolution 113* (1940), Part 15A, p. 8,791.

[10] A. J. Kraemer, "Developments in Petroleum Refining Technology in the United States, 1914–19," Information Circular, Department of Interior, Bureau of Mines (June, 1941).

[11] Polymerization, the complete reverse of cracking, utilizes lighter hydrocarbons and combines them into molecules of gasoline. See "Investigations of the Concentration of Economic Power," *loc. cit.*, Part 14, p. 7,479. John E. Shatford, in a statement made to the Committee on October 2, 1939, which appears in Part 15, pp. 8,147–64, said: "Another important factor in improving both the yield and quantity of gasoline by chemical means has been the development of polymerization processes which convert a part of the available refinery gases into a high anti-knock gasoline known as 'polymer' gasoline. The cracking process which breaks up large, high-boiling molecules into small, low-boiling molecules may be likened to the process which crushes lump coal to make the desired household

century, one barrel of Mid-Continent oil yielded the following products:[12]

AVERAGE REALIZATION AND COST FOR SMALL MID-CONTINENT REFINERS

	Per Cent Yield	Gallons from Barrel	Price per Gallon	Total Reali- zation
Gasoline—				
Premium	12.04	5.057	$0.10375	$0.5247
Regular	40.16	16.867	0.09625	1.6234
Kerosene	2.40	1.008	0.08250	0.0832
No. 1 distillate & furnace oil	9.60	4.032	0.08000	0.3226
No. 2 distillate & furnace oil	11.40	4.788	0.07375	0.3531
Diesel	2.00	0.840	0.08000	0.0672
Residual	16.60	6.972	0.03214	0.2241
Propane & gases	2.00	0.840	0.05500	0.0462
Loss	3.80	1.596	———	———
Total Realization	100.00	42.000	———	$3.2445

sizes. In both these processes there is the inevitable production of particles smaller than those desired, which take the form of fines in the crushing of coal, and of refinery gas in the production of gasoline. Briquetting processes have been developed in the coal industry to cement these small particles back into the more valuable sizes, and the polymerization processes likewise operate to force some of these small molecules to recombine into molecules of the right size and boiling point for gasoline. Fortunately, gasoline molecules formed by these processes have unusually high anti-knock value."

[12] *The Oil and Gas Journal,* Vol. XLVIII, No. 39 (February 2, 1950), 32. The yield of gasoline varies according to the grade of crude oil and the market value of other by-products. Pennsylvania crude, for example, is refined into superior grades of lubricating oils; it has a paraffin base, contains no sulphur, and has a lubricating content of 23 per cent, as compared to the usual 3 per cent in the crude oils of the United States. Manufacturers can refine nine gallons of lubricating oil from one barrel of Pennsylvania crude, while from one barrel of average United States crude oil, only one and one-tenth gallons can be refined.

According to *The Oil and Gas Journal,* Vol. XLVIII, No. 24 (October 20, 1949), 149, the oil-refining industry sold its products in 1926 for 160 per cent of the cost of crude oil, but during the first eight months of 1949, the price which refiners received for by-products had dropped to 105 per cent of the cost of the crude. Manufacturing, administrative, and selling expenses for the refining industry average about 22 per cent of the cost of crude oil.

Although industrial and domestic use of natural gas spread rapidly throughout the Middle West by the extension of pipe lines from the Mid-Continent area—its use was already common in Southern California, Pennsylvania, and West Virginia—it has been estimated that more than 1,250,000,000 cubic feet of gas were wasted each day from 1922 to 1934. Harold L. Ickes, secretary of the interior and at the time oil administrator under the provisions of the National Industrial Recovery Act, appeared at Dallas, Texas, on November 14, 1934, to speak before the annual meeting of the American Petroleum Institute. He included this statement:

> I had not expected to discuss at this time wasteful methods employed in the production of oil, but there has just been brought to my attention an example of such cruel and devilish waste that it almost staggers the imagination. There was brought to me in Washington, while I was working on this speech, a printed report entitled *Texas Panhandle Field, A Study of Gas Wastage and the Feasibility of Returning Waste Gas to Reservoirs,* which was prepared by an eminent committee of oil engineers whose names sponsor the report. I quote in part:
>
> "From reports in the Pampa office of the Texas Railroad Commission, we find that for the month of June, 1934, gas was being blown into the open air from the Texas Panhandle Field in the approximate daily amount of one billion cubic feet. The mere statement that one billion cubic feet is being wasted daily does not illustrate the significance of this great economic waste. It must be realized that the wastage is cumulative and continuous. The following comparisons tend to show what the wastage means:
>
> "1. The state of Texas ranks first in the United States in the sale of natural gas. Yet if the gas blown into the open air from the Texas Panhandle Field could be converted into gas produced and sold for a state which now has no gas at all, that

state would surpass all other states in the Union and run Texas a close race for first place.

"2. If 62,634,000 barrels of fuel oil were poured onto the ground, the waste of heat energy would be the same as the waste of one billion cubic feet of gas per day for a period of one year.

"3. Further significance of the magnitude of this waste is illustrated by the fact that in one year it amounts to the equivalent in heat energy of 24,333,000 tons of Texas lignite coal. This amount of coal would fill 487,000 freight cars. . . ."[13]

California, which has produced more oil than any state other than Texas, was the first to curb excessive production of gas from oil wells.[14] The Santa Fé Springs in Long Beach and the Huntington fields were producing more than two-thirds of the state's production in 1923, with little regard to the oil-gas ratio. Twelve oil and gas zones were exploited in the Santa Fé Springs field, varying in depth from 3,450 to 8,000 feet. On far-famed Hell's Half Acre wells were drilled to such a density that the legs of derricks interlocked; the field was divided into 1,145

[13] Ickes' speech appears in an appendix in Samuel B. Pettingill's book, *op. cit.*, 251–57.

A statement that the late J. C. Hunter of Abilene, Texas, at the time president of the Mid-Continent Oil and Gas Association, made to a Congressional committee on February 22, 1940, bears repeating: "It has been shown that the primary function of gas in an oil field is to act as a propelling force in the production of oil, and that the subsequent utilization for the manufacture of gasoline and the delivery for sale of residue gas or its reinjection into the reservoir are not always technically feasible or commercially possible." See Hearings before the Subcommittee of the Committee on Interstate and Foreign Commerce, 76 Cong., *House Resolutions 290 and 7372* (1940), Part 4, p. 1,962.

As late as November 27, 1948, the chairman of the Texas Railroad Commission issued a press release stating: "The sky is lighted in many parts of our State by oil-well gas flares burning valuable gas to the air. Every citizen has seen this waste going on. . . . Our program is to stop all such waste wherever the quantity is great enough to justify saving the gas." See "Oil Supply and Distribution Problems," 81 Cong., 1 sess., *Senate Report 25* (1949), 14.

[14] J. Howard Marshall, "Legal History of Oil and Gas in California," in *Legal History of Conservation of Oil and Gas,* 28–36.

parcels at the surface, many as small as 75 by 100 feet.[15] Before deeper zones were found, the Santa Fé Springs field, in June, 1923, produced 6,200 barrels of crude oil and 300,000,000 cubic feet of gas daily; a year later, production was 1,100 barrels and 50,000,000 cubic feet, respectively. Production of oil had declined with the decline of underground gas pressure. Since the same thing was happening in other fields, a serious study was made during the next four years which produced valuable data in regard to the oil-gas ratio. The California Legislature in 1929 amended a statute of 1915 to provide for a better definition of the unreasonable waste of gas. The state was given authority to determine whether unreasonable waste occurred in the production of crude oil and to establish ratios for any well or field. It was estimated that in the previous year 77,000,000,000 cubic feet of gas had been produced in the state through the operation of oil wells, little of which was converted to further use. In September, 1929, the Santa Fé Springs field was wasting 500,000,000 cubic feet a day. Oil-gas ratios were imposed, and court action sustained the state law.[16]

[15] The East Texas field, developed in the early 1930's, was the last great field where, in the townsites, derricks were stacked against one another. For example, at Kilgore, Texas, as well as at Triumph Hill, at Beaumont, and at Long Beach, one could see this phenomenon repeated from experience with flush production from the time of the Pennsylvania experience in the 1860's. At these places, holes were bored as closely as derrick-space would permit. At Kilgore, too, can be seen a producing oil well on the site of a razed bank building; the mezzanine floor was not removed and remains in place, with the exception of the space used for the well bore.

In 1927, the city of Oxford, Kansas, prepared and adopted a comprehensive ordinance regulating drilling permits within the city limits. It provided for regulatory well-spacing, and that owners should have an interest in benefits in proportion to their area of surface ownership. Similar city ordinances, when legally drawn, have withstood court action and provided relief from promiscuous drilling activity to residential areas throughout the country. See Innis D. Harris, "Legal History of Conservation of Oil and Gas in Kansas," in *Legal History of Conservation of Oil and Gas*, 55–56.

[16] *People* v. *Associated Oil Company et al.*, 211 Cal. App. 93 (December, 1930); *Bandini Petroleum Company* v. *Superior Court*, 110 Cal. App. 123 (November, 1930); and *Bandini Petroleum Company et al.* v. *Superior Court, Los Angeles, California, et al.*, 284 U. S. 8 (November, 1931).

Meantime, co-operative agreements had been in effect among operators in the Ventura field, California, for several years. For example, conservation measures caused the shutdown in 1927 of six wells, when it was found that the production of one barrel of crude oil released from 7,000 to 28,000 cubic feet of natural gas from the reservoir; in a seventh well the lowering of tubing resulted in the reduction of the oil-gas ratio; and another well that had been producing 2,000,000 cubic feet of gas daily and only 30 barrels of oil was redrilled.[17] Interest in voluntary co-operative agreements by oil producers prompted the American Petroleum Institute to conduct a special study in 1927 of the function of natural gas in the production of oil. This study revealed the importance of reservoir energy as the common property of all the owners of the mineral rights in a pool.[18] By the mid-thirties, it had become common practice to pass gas produced in oil-lift through a series of containers in which the pressure was reduced. Gasoline was extracted at each tank, and the natural gas was used to furnish gas-lift to older wells and to extract additional gasoline.

Because production of crude oil exceeded market demands in the late 1920's, proration—a movement to limit production from a field to a figure below capacity production—was inaugurated, first in the Seminole field in Oklahoma in 1926, then in the Yates Pool in Texas in 1927, next in the Hendricks Pool

[17] Miller, op. cit., 72.

[18] States were slow to adopt measures in effect in California, however; Oklahoma did so in 1933, Texas in 1935, Kansas in 1937, Arkansas in 1939, and Louisiana and New Mexico in 1940. In March, 1933, the Oklahoma Corporation Commission attempted to set an oil-gas ratio for the west Holdenville field, where some wells showed a ratio as high as 126,000 cubic feet of gas to one barrel of oil. The Commission set the ratio at one barrel of oil to 5,000 cubic feet of gas, but, on protest, rescinded the order. Claude Barrow, oil editor of the Oklahoma City Daily Oklahoman, on March 7, 1933, pointed out that "if operators take their allowable of 5,000 barrels a day there will be produced also 182,440,000 cubic feet of gas per day—enough to supply the entire state of Oklahoma. The 5,000 barrels of oil net producers $2,600 per day; the gas, at 6 cents per thousand feet, would net $10,946 per day. The loss due to lowered pressure can only be figured when they have to repressure their wells."

in Texas in 1928, and in the Hobbs field in New Mexico in 1930.[19] Oklahoma in 1928 and Texas in 1930 issued state-wide proration orders; Kansas did the same in 1931. At the present time, ten states limit production to market demand.

Experience gained in the Seminole field evolved into practices that led to stronger conservation legislation, court approval, and the adoption of state-wide proration by Oklahoma and other states. Flush production in the Seminole field began in July, 1926, when Fixico No. 1 blew in at more than 9,000 barrels of crude oil daily. Intensive development followed; by November 1, operators in the Mid-Continent area realized that overproduction and economic waste threatened the industry unless voluntary controls could be established at Seminole.[20]

In November 6, Ray M. Collins was selected by the oilmen to act as an umpire in the enforcement of self-imposed regulations adopted by the group. A committee of five was chosen to advise and assist Collins. Because of limited pipe-line facilities, the operators, fearing a waste of oil, agreed to curtail drilling activities, with the exception of direct offset wells, until November 25. Other wells were stopped at the top of the sand. This period of inactivity in drilling allowed the pipe-line companies enough time to extend and enlarge their lines and storage facilities in the field.

Production immediately increased after the period of the voluntary shutdown ended. There was no current need for the exploitation of the pool; the oil markets of the world were already being supplied with sufficient fuel to meet current de-

[19] Under proration, a well or lease is limited to a designated amount of the field's allowable production. Oklahoma first issued a proration order in 1915; this pertained to production from the Healdton field.

[20] The Seminole experience was chronicled in great detail in two contemporary accounts: See Swarts, Bapp, and Morris, "Petroleum Engineering Report on the Seminole Pool," *loc. cit.*, and Ray M. Collins, "The Mid-Continent Production Situation," a paper presented to the session of the General Committee of the Division of Development and Production, at the Ninth Annual Meeting of the Engineering Division, American Petroleum Institute, December, 1928.

mands. Producers in the area, however, forced hundreds of thousands of barrels of oil on an already saturated market. Seminole oil was reduced from $2.69 to $2.10 a barrel on November 17, and on March 12, 1927, when production reached 265,000 barrels a day, the price was further reduced to $1.28 a barrel.

Further curtailment became necessary when the Earlsboro and Bowlegs fields were opened as the first extensions north and south of Seminole. These later included nine pools, and they made the Greater Seminole area one of the most prolific of all times. Improved methods of production, particularly the application of the air-gas lift, contributed to an increased rate of recovery. The peak of production was reached on July 30, 1927, with 529,596 barrels produced daily from 666 wells.

As new pools were opened near Seminole, the jurisdiction of Collins and the advisory committee was extended to include the entire area. Most of the operators in these pools were major companies that owned pipe lines, tank cars, refineries, and market systems; and some of the independents found that, although they had contracts for their crude oil, pipe-line companies cut down their runs on the grounds, where their lines were crowded with production from other leases.

Some dissatisfaction was expressed because all the members of the committee of advisers were representatives of major companies. When the advisory committee was enlarged to seven members late in 1928, three were chosen from the ranks of the independents. Provision was also made for a legal committee of five lawyers to act as a consultative body. Collins continued as umpire, and the members of the Corporation Commission and the Attorney General represented the state.

This group worked out the details of the first proration order sanctioned by the state. The procedure worked out for the issuance of an order followed this plan: Discussion was held in open meetings at which operators were given an opportunity to offer suggestions; recommendations were made to the Cor-

179

poration Commission and referred to the Attorney General; the Commission would hold hearings for further consideration of the problems and requests presented and then issue an order usually in line with suggestions made by the advisory committee.

The pattern worked out by the operators for the Greater Seminole area served as the model in state-wide proration practices adopted later. Wildcat wells opening possible flush pools were pinched in at 100 barrels a day or shut down completely. Key wells were drilled, by agreement of producers, in certain areas to test the limits of the pool. The operator was paid two-thirds of the cost of a dry hole; if the well was a producer, it had to be pinched to 50 to 100 barrels a day until offsets were drilled. Only direct offsets were allowed, and shooting was permitted only in wells which were offset by wells that had already been shot.

An attempt was made to limit production in a certain area to a figure equal to the transportation and marketing facilities for its oil. The potential daily production for each lease or tract was determined by gauging the wells for a set number of hours; then, each lease was allowed to produce an amount equal to its percentage of the total allowed production. Many producers favored unitization of operations, but it was impossible to unitize every pool. Well-spacing differed in various pools. A concerted effort was made to conserve gas.

Difficulties arose to perplex Collins in this first attempt at prorating a great producing area. Many independents retained their distrust of integrated companies and had difficulty adjusting to a working relationship with the large companies; producers who had barely been meeting expenses in old pools wanted to "clean up" in the flush pools by unrestricted drilling and production; still others expressed fear that proration would be considered by federal authorities as a violation of the Sherman Antitrust Act; a few believed that by appealing to the Corporation Commission the oil industry was drifting toward governmental control.

Members of the legal staff of the committee had to consider such questions as the powers of the Corporation Commission, the rights of royalty owners to force the drilling of a well off-setting a pinched-in well, the attempts of royalty owners to compel the operators of their shut-in wells to open them to full production, the rights of lessees and lessors in deciding the ratio of potential production, and contracts on lease obligations which required either the drilling of a well within a specified time or the cancellation of the lease. These and many other problems were faced by the committee. Collins said later that the best results of the conservation movement in the Greater Seminole area were attained not through proration, but by agreements made by the producers to shut in producing wells or wells that looked as though they were on geological highs with great possibilities of opening new pools at a time when another flush pool would have spelled disaster to the industry.

Proration in Oklahoma was proving a success when the discovery well of the great Oklahoma City field was brought in early in December, 1928. Again, producers were faced with the possibility of opening another great reservoir of oil at a time when overproduction was already a problem. With the exception of a shutdown lasting for thirty days from September 11, 1929, operators raced one another in sinking wells with little regard to well-spacing and proper utilization of reservoir energy. The company that opened the field began with forty-acre spacing of wells, but other companies soon forced spacing down to ten acres, and early in 1930 drilling began on platted additions within the corporate limits of the city, on lots 50 feet by 140 feet.

Disregard of well-spacing and unitization led to the completion by 1931 of 765 wells, with a potential production twenty times that which could be marketed. Three hundred wells would have been adequate for a normal recovery from the Oklahoma City field, and the expenditure of $31,000,000 could have been saved.

181

New wells of large capacity were coming in almost daily in the Oklahoma City field during the early months of 1930. Since demand for oil from different pools in the state varied, the Corporation Commission issued an order on June 30 limiting production by varying percentages from field to field. The Oklahoma City field was limited to 8 1/3 per cent of its potential production.

But proration did not meet with universal favor among oilmen. There were some who practiced the most up-to-date methods of conservation in producing crude oil, but balked at curtailment; integrated companies denied they were guilty of wasteful methods of production; companies with holdings in other states thought it inequitable to enforce proration in Oklahoma while elsewhere flush production was permitted and a flood of foreign imports entered. Doubt of the Corporation Commission's authority to act as an administrative body with legislative and quasi-judicial powers was expressed by Chief Justice Fletcher Riley of the Supreme Court of Oklahoma in a dissenting opinion on February 16, 1933. He added these words: "In my opinion, proration of oil was born of monopoly, sired by arbitrary power, and its progeny (such as these orders) is the deformed child whose playmates are graft, theft, bribery and corruption."[21]

When in August, 1930, the operators' committee and the umpire ordered the C. C. Julian Oil and Royalties Company to shut in a well in the Oklahoma City field for sixty-five days and then produce it only one day out of twelve, the company sued the Corporation Commission for a writ of prohibition. They contended that the Conservation Act of 1915 was unconstitutional in many of its provisions; that the act itself, even if valid, did not confer the authority attempted to be exercised; that the orders of the Corporation Commission went far beyond the pro-

[21] *H. F. Wilcox Oil and Gas Company* v. *State et al.,* 162 Oklahoma 89 (February, 1933).

visions of the law; and that since the act forbade unreasonable discrimination in favor of one common source of supply as against another, the Corporation Commission's orders issued on June 30 and July 24 were void.[22]

The Julian Company held oil and gas leases on three 25-foot lots, an area 75 by 140 feet, and had drilled a well capable of producing 5,000 barrels of oil a day. The company argued there was a market demand at the well for the oil and that there would be no waste since the oil would be piped to its near-by refinery. Nor did the company think it right that "16 other oil fields in the state are allowed to produce 50 per cent of their potential production, two fields 25 per cent, and five fields 18¾ per cent, and that many other fields in the state are not prorated at all."

The Oklahoma Supreme Court sustained the Conservation Act and the authority exercised by the Corporation Commission. An illustration of how far the public mind and court decisions had departed from the plundering doctrine expressed by the rule of capture is found in the opinion: "Surface owners of land have the right to drill for and reduce to possession the oil and gas beneath; but this right is to all of the owners alike; and when numerous surface owners seek to produce from a common pool, it is within the police power of the state, in keeping with due process of law, to require the several surface owners to produce same under reasonable regulations to the end that some of said owners may not take from the common source more than their equitable share."[23] Later, in a testing of the Oklahoma proration law in the federal courts, the statute was upheld.[24]

Proration of the Oklahoma City field, nevertheless, suffered

---

[22] The Corporation Commission Order of July 24, 1930, Order No. 5246, reduced the state allowable to 550,000 barrels daily. This gave the Oklahoma City field an allowable of 18¾ per cent, but this was reduced to 8⅓ per cent on July 30.

[23] C. C. Julian Oil and Royalties Company v. Capshaw et al., 145 Oklahoma 237 (October, 1930).

[24] Champlin Refining Company v. Corporation Commission of Oklahoma, 286 U. S. 210 (May, 1932).

a setback from a decision of the state Supreme Court in February, 1933. This was the result of an action of the Corporation Commission against the H. F. Wilcox Oil and Gas Company for its refusal to make daily production reports and on the charge that it had produced 1,249,843 barrels of oil above its prorated amount. Ordered to close its ten wells in the Oklahoma City field until the Commission should be shown that the overproduction was made up, the Wilcox Company took exception to the order because it applied proration to the Oklahoma City field as the common source of supply and not to the separate formations from which oil is produced. The company showed that there were four common sources of supply of natural reservoirs from which wells were producing, and the Supreme Court held that the Commission was not authorized to prorate the total market demand from four separate sources, but that its authority was limited to proration of the market demand among the wells producing from a single source of supply.

The Oklahoma Legislature revised the conservation law. This statute, signed by the Governor on April 10, 1933, contained forty-nine sections and listed in detail powers delegated to the Corporation Commission for enforcement of its provisions. The definition of what constitutes waste according to 1915 standards was amended to include water encroachment in oil and gas strata. Conservation of gas was provided for, and the importance of reservoir energy was stressed. A system of reports on every well and a checkup on the movement of oil was provided for in an effort to stop the illegal production and transportation of oil. A better system of taking well potentials was written into the act, and severe penalties were provided for violations of orders issued by the Commission. The offices of proration umpire, assistant umpire, and proration attorney, all under the Conservation Department of the Corporation Commission, were created.

Orders for the Oklahoma City field, which long had been the source of most of the irregularities in enforcement of conservation practices and proration measures, were immediately prepared and put in operation. The different sources of supply were recognized, and a more orderly method of deriving the potentials of wells was instituted. A comprehensive order was issued in mid-summer which classified, for the purpose of proration, all the pools in Oklahoma into one of four classes. This method of procedure has since been followed, and each month the Commission holds an open hearing before establishing proration figures for each pool on a well or lease basis.

The Conservation Act was amended in 1935 to provide for well-spacing upon the development of new pools and to include a stricter definition of gas waste as it affects reservoir pressure. Operators had witnessed the efficient manner in which the South Burbank field on Osage Indian land was being produced under federal regulations that limited the number of wells and conserved reservoir energy. When the prolific Fitts field came in, in September 1933, it had been placed under a voluntary 10-acre spacing restriction, with each well placed in the center of the tract. Following the operator's lead, the 1935 statute provided that the drilling unit should not exceed 10 acres unless 80 per cent of the lessees owning at least 80 per cent of the tract requested wider spacing, but, in any case, the drilling unit should not exceed 40 acres.[25] On the opening of each new pool, the Corporation Commission decided the required well-spacing.

This law was immediately attacked, but was sustained by the United States Supreme Court.[26] The Secretary of the Interior, Harold L. Ickes, under emergency powers granted in 1933, is-

[25] This act was amended by legislative enactments in 1945 and 1947 to provide that the Corporation Commission could establish well-spacing without the percentage of ownership consent. Spacing units of 40 acres are sometimes established for production from deeper pools.

[26] The Supreme Court of Oklahoma upheld the well-spacing act; the appeal to the United States Supreme Court was dismissed. See *Patterson* v. *Stanolind Oil and Gas Company,* 182 Oklahoma 155 (March, 1938); 305 U. S. 376 (January, 1939).

sued an order for 40-acre spacing in developing oil reserves, but the following year a federal District Court held that this order was unauthorized.[27] From December 23, 1941, to September 1, 1945, however, under emergency war powers and because of critical shortages in materials, 40-acre spacing was reinstated.

Limitation of waste from the production of crude oil—or its overproduction—was fiercely fought in legislative halls and state and federal courts during the 1930's. Texas, the greatest producing area, and Oklahoma, the pioneer in conservation measures, were the scenes of greatest struggle and triumph. Following educational campaigns by the industry through the Interstate Oil and Gas Compact Commission, the research of petroleum engineers, and comprehensive studies and open hearings by legislative committees, state after state revised laws relating to waste in oil production or enacted more adequate conservation measures. The state laws have been subjected to attack and change; adopted amendments, however, have usually strengthened recommended conservation practices.[28] Limitation of production to market demand is written into the statutes of ten states; greater knowledge of the importance of reservoir pressure brought adoption by most of the states of legislation similar to that of California, which limited the oil-gas ratio in production; well-spacing and unitization of production have gained favor; and depth of well, acreage, or bottom-hole pressure have become factors in limiting production.[29] Practically every prob-

---

[27] *United States* v. *Eason Oil Company,* 79 Fed. (2d) 1013 (June, 1935). The federal government and nine states—Alabama, Arizona, Arkansas, Florida, Georgia, Illinois, Indiana, Louisiana, and Oklahoma—had passed acts by 1948 to provide for compulsory unit operations under certain circumstances, and six states—California, Kansas, Michigan, New Mexico, Texas, and Wyoming—permitted voluntary agreements. Unitization is necessary in secondary recovery operations.

[28] The Louisiana Conservation Act of 1940 became a model in regard to the elimination of waste in oil production. The Legal Committee of the Interstate Oil Compact Commission has rendered invaluable assistance to states on proposed or recommended legislation.

[29] Limitation of production to market demand can adversely affect one pool

Barrel factory on Oil Creek (Drake Museum).

24. Wooden storage tanks in Pennsylvania in 1868 (Standard Oil Company, N.J.).

. Three sizes of barrels on platform of railroad depot, *bottom left* (Drake Museum).

. Storage space for an old pumping well, *bottom right* (Standard Oil Company, N.J.).

27. Cannon used to pierce steel sides of early oil tanks to relieve pressure within and prevent burning of hot oil over the fire walls (Cities Service Company).

28. A modern oil refinery (Cities Service Company).

29. Section of a pipe line laid in 1949 (Cities Service Company; photograph by Fritz Henle).

30. Loading oil for shipment on a tank car (Cities Service Company).

Modern drilling mast (*The Oil and Gas Journal*).

33. A 204-foot oil derrick (Shell Oil Company).

A roughneck on the floor of a drilling rig (Cities Service Company; photograph by Fritz Henle).

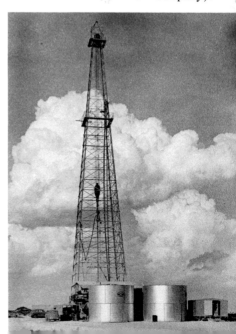

34. Helicopter and barge unloading supplies at an offshore rig, *top left* (Vertol Aircraft Corporation).

35. Offshore rig with quarters, storerooms, storage tanks, and other facilities maintaining the well, *top right* (Standard Oil Company, N.J.).

36. Substructure of a new offshore rig.

urveyor exploring for new oil resources in dry, rocky hills (Cities Service
ompany; photograph by Fritz Henle).

est well drilled atop a 2,000-foot plateau
Utah (Cities Service Company; photo-
aph by Fritz Henle).

39. A geologist searches for oil on
    sand dunes of Texas (Cities
    Service Company; photograph
    by N. J. Parrino).

40. Seismographic party
swamp buggies and pirogu
conduct their search for
bearing formations under
swamp beds (Cities Ser
Company; photograp
Fritz Henle).

41. A seismograph operator uses geophones to pick up seismic waves reflected
from near-by shot, *bottom left (The Oil and Gas Journal)*.

42. Seismographic exploration off the Gulf of Mexico, *bottom right* (Cities Service Company; photograph by Fritz Henle).

3–44. Superhighways and bridges reflect the progress made possible by modern motor fuels and lubricants (Cities Service Company; photographs by C. E. Rotkin).

45. An attendant at a retail gasoline filling station (Cities Service Company; photograph by Nelson Morris).

46. Pure parrafin wax is extracted oil (Cities Service Company; p graph by Fritz Henle).

48. A housewife cooks with purifie ural gas (Cities Service Com photograph by Fritz Henle).

47. The demands of high-powered aircraft engines are met by oil products (Cities Service Company; photograph by Tony Linck).

lem relating to production can be solved except the "stripper"-well problem.

When voluntary proration began in 1926, no one thought this form of conservation would reach out to include marginal or "stripper" wells, which produced from a fraction of a barrel to ten barrels of oil daily. The life of such a well—whose production expense approximates income—is indeterminate; the Colby well, drilled in August, 1861, near Oil Creek in Pennsylvania is still producing about one-half barrel of crude oil a week.[30] Oil wells are usually abandoned when their production will no longer cover operating costs; the selling price of oil, therefore, is a major consideration in determining when a well will be abandoned.

Despite the opening of new fields each year, most of the production of crude oil in the United States comes from wells which produce twenty barrels or less each day. States in the Mid-Continent area, while experimenting with proration and shut-down orders in the 1930's, quickly learned the disastrous effects of such controls upon stripper-well production. Testing made by the Bureau of Mines proved that well casings corrode, in some areas at least, three times as fast in shut-in wells as in producing wells. Severe corrosion may cause the casing to collapse and thus necessitate redrilling before the well may again be pumped. Premature abandonment of wells may involve seepage of surface and underground waters into the oil-bearing stratum and result in a substantial hidden loss in ultimate recovery. Since 1878, when Pennsylvania enacted the first law,

or region in comparison with another. Kansas, for example, has suffered because Oklahoma and Texas have historically claimed a disproportionate share of the market; Texas claims her Gulf outlets, and Oklahoma, her pipe-line and refinery connections. Although production increased tenfold in Kansas from 1933 to 1938, its allowable based upon market demand remained unchanged. See "Kansas, 1937–48," in Murphy (editor), *op. cit.,* 160–62.

[30] "Petroleum Study," in Hearings before a Subcommittee of the Committee on Interstate and Foreign Commerce, 81 Cong., *House Resolutions 107 and 6047; House Joint Resolution 423* (1950), 24.

state regulations have been very strict in regard to plugging abandoned wells, and most operators in old fields have zealously guarded their rights to produce from stripper wells as long as possible.

Pipe-line carriers of crude oil or any purchasing company can connect with a very few wells in a productive field and receive an equal or greater volume of crude oil at a minimum gathering expense compared with the greater expense of gathering from a large number of small producing wells in a marginal field. The National Stripper Well Association during the 1930's faced crisis after crisis because of this basic economic fact, which is as old as the beginning of flush production along Oil Creek in 1861, and it is only recently that some of the oil-producing states have been successful in keeping purchasing companies in the stripper fields.

With the outbreak of World War II, the equipment of thousands of marginal wells became worth more than the oil they produced. Some of the owners agreed with the opinion expressed in 1942 by Roy M. Johnson, an Oklahoma oilman, when he was quoted as stating he had been offered $3,500 for a stripper well, "but I'll be darned if I'm going to sell that well for junk." Many operators, nevertheless, sold stripper-well equipment for its salvage value. Because of the war emergency, the Office of Price Administration in August, 1944, established a subsidy of thirty-five cents a barrel for crude oil from fields with an average production per well of less than five barrels daily, and twenty cents a barrel for fields averaging five to nine barrels per well each day.

11

CONSERVATION

THROUGH CO-OPERATION

IN THE EARLY 1930's, operators realized that state action alone could not achieve economic conservation. Wasteful methods and overproduction in states unhampered by regulatory restrictions had to be checked. Individual states had proved their power to regulate the production of oil and gas within their borders. Laws had been passed to restrict the blowing off of gas and the quantity of gas producible per barrel of oil. In some states, a limit had been placed upon the quantity of production from a field in accordance with some standard of waste, such as the limitation of production in excess of market demand or the prevention of physical waste. Allowable production could be allocated among the wells of a field in proportion to their potential production or at a flat rate per well. Some attempted to require that pipe lines become common purchasers and buy from all producers ratably. Proration of production had been established not only among the wells of a single pool but also among the competing fields of a state. The United States Supreme Court had sustained the Indiana gas-waste act, the Wyoming carbon-black act, the California gas-oil ratio law, and the proration statute of Oklahoma.[1]

[1] *Ohio Oil Company v. Indiana,* 177 U. S. 190 (April, 1900); *Walls v. Midland Carbon Company,* 254 U. S. 300 (December, 1920); *Bandini Petroleum Company et al. v. Superior Court, Los Angeles, California, et al.,* 284 U. S. 8, (November, 1931); *Champlin Refining Company v. Corporation Commission of Oklahoma et al.,* 286 U. S. 210 (May, 1932). See also *State and Federal Conser-*

For almost ten years, leaders in the Mid-Continent area had concluded that conservation practices should not be limited by state lines, although there was the reasonably expressed fear by others that restraints leading to curtailment would be considered a violation of the Sherman Antitrust Act. E. W. Marland, a Mid-Continent producer and refiner, declared at a meeting of the Western Refiners Association held in Kansas City on July 21, 1923, that producers and refiners must overcome the effects of overproduction either through the "ruthless and inexorable law of supply and demand working through the price factor, or by intelligent curtailment of output."[2] The meeting endorsed Marland's statement, but did so with the expressed reservation that no action be taken.

Some of the producers were publicly advocating regulation the following year, when flush production in the Mid-Continent area continued to add to the oversupply. H. H. Gray, president of the National Organization of Independent Producers, advocated the limitation of production by the curtailment of exploratory activity. He urged that all fields of flush production be placed under some system of proration. A few weeks later Henry L. Doherty, the head of a big integrated company, proposed measures for stabilization of the oil industry which included the issuance of permits by the state before wildcat drilling, viz., drilling in unproved territory, could be undertaken.

Representatives of two influential organizations of the industry—the Mid-Continent Oil and Gas Association, organized in 1917, and the American Petroleum Institute, founded in 1919, which became the greatest single agency sponsoring the interests of the oil industry—issued reports on the need of stabilization.

A federal Oil Conservation Board was created under Presi-

---

vation *Laws and Regulations Relating to the Production of Oil and Gas,* issued by the Federal Oil Conservation Board; also, *Report V of the Federal Oil Conservation Board to the President of the United States* (October, 1932).

[2] Oklahoma City *Daily Oklahoman,* August 1, 1923.

dent Coolidge in 1924, and members of the industry attended meetings with the federal officials to discuss problems peculiar to the industry. Generally, the industry favored voluntary regulation over governmental interference. The cautious attitude toward federal regulation was expressed by Charles Evans Hughes, a representative of the A.P.I. (American Petroleum Institute), in a public hearing before the Board on May 27, 1926: "The Government of the United States is one of enumerated powers and is not at liberty to control the internal affairs of the states, respectively, such as production within the states, through assertions by Congress of a desire either to provide for the common defense or to promote the general welfare."

A contrary view was expressed by Henry L. Doherty of the Cities Service Oil Company: "If the Federal Government has no power to conserve oil and prevent waste, then our plan of government is defective, because the power is not vested any place for us to do that which may be necessary for our national defense." During the next few years producers made general suggestions to the federal Oil Conservation Board, without conceding any of their rights to conduct their business in their own independent fashion. Evidently they were afraid that government interference would lead to undue regulation.

This remained the general attitude of producers until the market crash of 1929 and the severe depression following led them to ask for federal assistance. Hubert Work, the secretary of the interior, suggested that governors of oil-producing states should co-operate in an effort to secure uniform state legislation for the practical conservation of the country's natural resources.

When Work's successor, Ray Lyman Wilbur, called a meeting of governors or their representatives from the various oil-producing states at Colorado Springs in June, 1929, the idea of oil conservation through interstate agreement was discussed, but there was unanimous objection to the creation of an interstate compact. It was the opinion of the meeting that state legislatures would not sanction such a compact, and that the oil industry, at least in Oklahoma, and probably in Texas and California, "would never consent to a super-commission of this nature."

The request for co-operation among the oil-producing states was renewed by Secretary Wilbur. Oklahoma and Texas were each producing more petroleum than the three leading foreign countries—Venezuela, Russia, and Mexico—combined, and the enormous production of the East Texas area was a serious threat to stability. Governor William H. Murray of Oklahoma urged the governors of states in the Mid-Continent group to consult and advise together on means of correcting the evils of over-production. The governors of Texas and Oklahoma met with representatives of the governors of New Mexico and Kansas at Fort Worth in February, 1931, to perfect plans for an Oil States Advisory Committee. This committee appeared before the federal board in Washington on April 9, 1931, and presented the following suggestions:

1. Stabilization of the production of crude oil is necessary to the stability of public supply and to the elimination of waste of an irreplaceable resource.

2. No individual oil producing state by its own laws can adequately protect the national interest—unless the conservation efforts of the large producing states shall be co-ordinated, otherwise curtailment efforts in one state may be offset and nullified by increased flush production in another, or by unjustified increases of importation to the great damage of the areas of old and settled production throughout the country.

3. With consideration duly given to the general situation of the oil industry throughout the United States, the stabilization of the production of crude oil within the state is purely a problem for solution by that state and the industry therein, with such cooperation and advice as may be rendered by interstate advisory committees, and with such aid and assistance as the Federal Government may be able to give.

Specifically, the Advisory Committee wanted limitation of unnecessary drilling, conservation of gas energy, "unitization" of pools, ratable takings within a given field, and an equitable apportionment of outlets from producing areas. A strong plea was made for greater uniformity and the stricter enforcement of state laws. To achieve this, the committee recommended that the legislatures of each of the oil-producing states authorize the negotiation of interstate agreements for the co-ordination of conservation measures.

Secretary Wilbur said that while the federal Oil Conservation Board was not empowered to make regulations for the oil industry, it did approve the suggestions for co-operation among the states. He promised that the board would continue making periodic examinations into the status of the oil industry and would issue regional and national forecasts of supply and demand. He endorsed the request that the states should have uniform conservation laws on certain major points, such as "unit" operation, protection against waste caused by overproduction, and conservation of gas energy.

Texas and California, in the meantime, were being accused of flooding the market. A California mineral conservation law aimed solely at the prevention of physical waste of oil and natural gas was tested and upheld in California courts. Wasteful storage of oil was prohibited, and companies were not allowed to let gas flow off into the air. Texas had made some efforts at conservation and, in the Yates Pool in Pecos County, voluntarily organized proration was begun in 1927. Although this was done with the consent and sanction of the Railroad Commission of Texas, which is the conservation agency of that state, there was no state conservation law, and by 1931 the great East Texas field was in flush production. At the same time, the Kansas Legislature was considering legislation similar to that in Oklahoma, while some Oklahomans were advocating California's type of law, and California was studying the conservation practices of the Mid-Continent states. All of the oil-producing states were interested in some form of joint action that would control overproduction and its depressing effect upon the price of petroleum products.

The seriousness of the world-wide depression caused leaders of all of the nation's industries to seek federal assistance. Three days after President Franklin D. Roosevelt took office on March 4, 1933, leaders of the oil industry appealed to Harold L. Ickes, the secretary of the interior, for a program of drastic regulation. He immediately called a conference, which was held before the end of the month. Fifteen governors and other special representatives of the oil-producing states, as well as many oilmen, attended. They agreed with Ickes that the shipment of "hot oil" could be controlled by the use of federal authority over interstate commerce. *The Oil and Gas Journal,* in its March 30, 1933, issue, quoted Wirt Franklin, then president of the Independent Producers Association: "This is the first time I have ever known of a unanimous report in the oil industry. You know how a man feels with the toothache. He will take anything to stop it. That is the way they are now."

Late in May, a committee of operators drafted a bill which was introduced in the House of Representatives by E. W. Marland, congressman from Oklahoma. This proposal, known as the Marland-Cooper-Ickes Bill, gave the secretary of the interior the power to fix prices, wages, and hours of labor, to limit production to demand, and to control the importation of oil. Although this bill had the President's approval, Congress was considering the $3,000,000,000 Industrial Recovery Bill, and it was decided to include the oil bill in the National Industrial Recovery Act. This was done, and the act, when passed on June 11, 1933, provided for reasonable pipe-line charges, ordered separation of pipe lines from holding companies, and forbade the transportation of "hot oil." In keeping with the general plan of the NIRA, a code was adopted for the oil industry which provided that the President should establish prices and regulate production. The drilling of new wells without federal consent was prohibited. On July 12, the President issued an executive order against the transportation in interstate commerce of illegally produced or "hot oil"; monthly reports were required of every shipment, as well as its source and time of production, the average flow of the wells, and other pertinent information which would enable a conservation officer to make an effective check. A division of investigation was set up to check reports and violations.

That these remedial measures were effective is borne out by Claude Barrow, who stated in the Oklahoma City *Daily Oklahoman* on October 15, 1933, that within two years, in excess of 15,000,000 barrels of crude oil had been illegally produced and sold from the Oklahoma City field, but that the regulatory measures enforced by the federal government had put an absolute stop to this type of oil thievery. The traffic in "hot oil" was stopped. When the price of crude oil rose from 25 cents in May, 1933, to $1.08 in October, 1933, oil producers were willing to credit part of the rise to the enforcement of the federal regulations. The authoritative *Oil and Gas Journal* pointed out on October 26, 1933, that "conservation cannot be attained if the price of crude oil and the products is so low that through competition with flush producing areas it becomes economically impracticable to operate the '300,000 stripper' wells in the nation, thus forcing their premature abandonment and the loss of billions of barrels of reserves."

Senator Elmer Thomas of Oklahoma introduced a bill in Congress on April 30, 1934, for unrestricted and permanent control over the oil industry by the federal government. A special Congressional subcommittee was appointed by the House of Representatives to visit the oil-producing states, to hold hearings there and to report to Congress on their attitude toward federal control. Successive sessions of Congress renewed the Pe-

195

troleum Investigation Committee. Opinions presented in the hearings before the committee in 1934 were divided. The cleavage between the independents and the major companies again asserted itself. The I.P.A.A. (Independent Petroleum Association of America) pledged support to Secretary Ickes, with the desire to be as helpful as possible to him in his important and difficult tasks, but the majors opposed further regulation. Governor Alfred Landon of Kansas praised the results of co-operation between the national and state governments. Governor-elect James V. Allred of Texas saw no necessity for giving the federal government the power and duty of controlling oil production in that state, while R. W. Fair, a member of the East Texas Tender Committee of the Railroad Commission of Texas, said that state agencies "are incapable of handling the entire problem, and the need is for Federal control just as strong as you can make it." Norman L. Meyers, executive secretary of the Petroleum Administration Board and chairman of the Federal Tender Committee, favored "legislation that would permit the Federal Government to control oil down to the individual well, if necessary."

Something more than an advisory commission was needed, however. The Independent Petroleum Association, with members from all producing states, met in Oklahoma City in October, 1934, and resolved that "excessive production causes surface, underground and economic waste; results in the rapid depletion of flush fields, and forces the premature abandonment of wells of settled production." The Association favored restrictions on imports, federal allocation according to market demand to the states, and, if any state exceeded its quota, federal control and regulation of shipments within the state. It also favored the planned, orderly development of new pools through agreement of the majority of operators, owners, and members in the area. The Association favored authorization by Congress of an interstate compact.

In November, the American Petroleum Institute held its annual meeting at Dallas, where Secretary Ickes appeared as the featured speaker. In blunt, incisive fashion, Ickes related how, the year before, a code had been established by the petroleum industry, by which it could govern itself with the co-operation of the federal government under the NIRA. He traced the effectiveness of interstate controls over the shipment of oil; he remarked on the continued depletion of reservoir energy and gas wastage in certain fields; and he implied that the industry should clean up its wastefulness of nature's resources, under the threat of stricter federal control.

Ickes left the convention divided in sentiment. The following day, however, a majority report was adopted by A.P.I. which advocated that the federal government recommend domestic allocations, continue to prevent interstate shipment of "hot oil," and regulate the importation of foreign-produced oil. The report also recommended that Congress be petitioned to give its consent to a compact among oil-producing states.

E. W. Marland, the governor-elect of Oklahoma, called meetings of the governors of the Mid-Continent area and their representatives in December and January, 1935, to discuss the feasibility of forming an oil-states compact. Marland pointed out that proration practices and conservation measures enforced in only one state discriminated against the producers of that area unless equitable restrictions were adopted in other oil-producing states. Representatives from the other states, too, were aware of the need for interstate co-operation, but they favored something short of federal regulation. The compact idea gained momentum as apprehension in regard to federal interference in the industry increased. The Constitution of the United States requires that compacts made between states must be ratified by Congress, and members of Congress from the oil states were prepared to push ratification of the proposed compact. On February 16, 1935, at Dallas, Texas, the compact was executed

197

by the representatives of the states of Oklahoma, California, Texas, and New Mexico, and was recommended for ratification by the representatives of Arkansas, Colorado, Illinois, Kansas, and Michigan.[3] Congress approved the compact on August 27.

The compact was to expire on June 1, 1937, but later Congresses have extended its life—first, over two-year periods; later, for four years. Hearings are held before a Congressional committee, however, before Congress acts to extend the compact.[4]

Governor Allan Shivers of Texas, at the time chairman of the Interstate Oil Compact Commission, summarized the accomplishments to a Congressional committee on June 11, 1951.[5] He stated that, since the inception of the compact, all member-states had improved or enacted comprehensive oil and gas conservation laws, that there had been an interchange of information relative to conservancy practices among the states and industry, that valuable pioneer work had been done on secondary recovery practices, and that progress had been made in the utili-

[3] The legislature of California still has not ratified the agreement; therefore, California is the only major oil-producing state that is not a member. The governor of California, nevertheless, usually sends an "observer" to Compact meetings. See Marshall, *op. cit.*, 28–36.

The following schedule shows when states became members or associate members of the Interstate Oil Compact Commission. (Associate members, those states which do not produce oil in commercial quantities, are marked with an asterisk.) Colorado, Illinois, Kansas, New Mexico, Oklahoma, and Texas (1935); Michigan (1939); Arkansas, Louisiana, New York, and Pennsylvania (1941); Kentucky (1942); Ohio (1943); Alabama, Florida, Montana, and West Virginia (1945); Georgia* (1946); Indiana and Tennessee (1947); Mississippi (1948); Arizona (1951); Nevada, Nebraska, North Dakota, and Washington* (1953); Oregon* (1954); South Dakota and Wyoming (1955); and Utah (1957). Alaska also became an associate member in 1953.

[4] The central office of the Compact Commission was established in the State Capitol Building in Oklahoma City. A modernistic building near by was dedicated in 1954 as permanent headquarters for the central office, which houses the executive secretary, the staff, the library, etc. Judge Earl Foster, executive secretary since 1944, devotes full time and energy to his administrative duties. His title was changed to General Counsel in 1957.

[5] "Interstate Oil and Gas Compact," Hearings before the Committee on Interstate and Foreign Commerce, 82 Cong., *House Joint Resolutions 206 and 211* (1951), 14.

zation of casing-head gasoline. He pointed out how effectively the Compact Commission had aided Congress and interested federal agencies in the gathering of information and the handling of petroleum problems connected with the defense effort during World War II and the Korean episode. He added: "The Compact States have demonstrated a clear understanding of their responsibilities in the field of oil and gas conservation, and a willingness to accept those responsibilities, while keeping uppermost in mind the public interest."

Compact Commission activities are financed by voluntary contributions from member-states. Each state contributes as much or as little as it deems appropriate, but no payment is necessary in order for a state to become a member. No outside contributions are accepted.

Until 1950 the Compact Commission held quarterly meetings each year; now it meets semiannually. At the December meeting, elective officers are selected—a chairman, a first vice-chairman, and a second vice-chairman. Since 1948 a governor of an oil-producing state has always been honored with the chairmanship.[6]

[6] Chairmen have been: Governor E. W. Marland of Oklahoma, 1935–36; Ernest O. Thompson of the Texas Railroad Commission, 1936–40; Governor Leon C. Phillips of Oklahoma, 1940–42; Governor Andrew F. Schoeppel of Kansas, 1942–45; Governor Robert S. Kerr of Oklahoma, 1945–46; Hiram M. Dow of Roswell, New Mexico, 1946–47; and Governors Beauford H. Jester of Texas, 1947–48; Frank Carlson of Kansas, 1949; Roy J. Turner of Oklahoma, 1950; Allan Shivers of Texas, 1951; Sid McMath of Arkansas, 1952; Edward F. Arn of Kansas, 1953; Johnston Murray of Oklahoma, 1954; William G. Stratton of Illinois, 1955; John F. Simms of New Mexico, 1956; Raymond Gary of Oklahoma, 1957; and Milward L. Simpson of Wyoming, 1958.

Ernest O. Thompson of Amarillo has been a member of the Texas Railroad Commission since June 1932, and no other man in the history of public service in the United States has contributed more toward the stabilization of oil conservancy practices than he. He began his long career of service on the regulatory body at a time when Texas oil could easily supply national domestic demands; he used his influence with governors and legislatures for stricter legislation. The American Petroleum Institute on November 7, 1951, made the following citation:

"The American Petroleum Institute Gold Medal for Distinguished Achieve-

In accordance with the bylaws of the Compact Commission, committees have been appointed to cover various phases of conservation. The Engineering Committee keeps the Compact Commission informed on the best and latest engineering and production methods for the orderly, nonwasteful production of oil and gas. A recent publication was a 128-page booklet, *Oil and Gas Production,* an introductory guide to production techniques and conservation methods. The Research and Co-ordinating Committee, composed of engineers and geologists from the various state regulatory bodies, keeps the Compact Commission informed of state projects that may be of interest and benefit to all the members. A Regulatory Practice Committee is composed primarily of administrators from state regulatory bodies, and makes reports on regulatory matters pertinent to the proper administration of conservation laws. The Legal Committee keeps

ment is hereby awarded to Ernest Othmer Thompson; Lieutenant-General; Senior Member of the Texas Railroad Commission.

"You pioneered in the conservation of oil and gas through the prevention of physical waste.

"You took affirmative leadership in the establishment and application of conservation statutes.

"You were one of the founders and three times Chairman of the Interstate Oil and Gas Compact to prevent waste in the production of oil and gas.

"You have been an active citizen-soldier all your mature life, and brought to your oil and gas regulatory work a recognition of the need for adequate daily reserve producing ability for defense.

"You made a worldwide oil study and on-the-ground survey in World War II for the Secretary of War.

"Through constant public utterances you were able to cause to be established by the University of Texas and the Agricultural and Mechanical College of Texas a permanent research institute to find ways to recover hidden oil in old thought-to-be depleted fields.

"So the American Petroleum Institute honors itself in paying this tribute to you."

James A. Clark's *Three Stars For the Colonel* is an excellent account of Thompson's life and achievements.

Hiram M. Dow, chairman of the Compact Commission from 1946 to 1947, took an active part in the preliminary meetings which led to the Compact and has continuously represented New Mexico on the Commission. (New Mexico was the first state to adopt the Compact.) He was chairman, during World War II, of the National Conference of Petroleum Regulatory Authorities.

the Commission advised on all legal matters affecting the conservation of oil and gas. This committee has prepared model conservation statutes which have been adopted by more recent members of the Compact. The Secondary Recovery Advisory Committee works through the secondary-recovery division in the headquarters office in compiling information on this subject.

Article V of the Compact is recurrently subject to scrutiny by Congressional inquiry, whenever the Compact is being considered for renewal or some phase of the oil industry is under investigation.[7] It provides: "This compact does not authorize the States joining herein to limit the production of oil or gas for the purpose of stabilizing or fixing the price thereof, or for the purpose of balancing supply and demand, or to create or perpetuate monopoly, or to promote regimentation, but is limited to the sole purpose of conserving oil and gas and preventing the avoidable waste thereof within reasonable limitations."

This provision obviously represents a compromise which was formulated when representatives of the states met to consider joining in a compact. Governor E. W. Marland of Oklahoma foresaw the need to limit production in order to obtain a better price for crude oil; Governor James V. Allred of Texas insisted that the sole purpose of a compact agreement should be the conservation of oil and gas by the prevention of physical waste.[8]

Ten of the oil-producing states have enacted laws which specifically authorize a limitation of production to reasonable market demand.[9] Federal and state courts have decreed that a state

[7] "Interstate Oil and Gas Compact," *loc. cit.;* also, "Petroleum Study," *loc. cit.;* and "Oil Supply and Distribution Problems," *loc. cit.* The latter is a report prepared by a special committee to study the problems of American small businesses. Often referred to as the *Wherry Report*—the chairman of the committee was the late Senator Kenneth S. Wherry of Nebraska—the report sharply criticized many practices of the oil industry.

[8] See "The Formation of the Interstate Compact to Conserve Oil and Gas," in Murphy (editor), *op cit.,* 556–70.

[9] Limitations of production to market demand appear in the statutes of the following states: Alabama, Arkansas, Florida, Kansas, Louisiana, Michigan, New Mexico, North Dakota, Oklahoma, and Texas.

may exercise its police power to prevent physical waste. This aspect of conservation, nevertheless, was singled out by the *Wherry Report* for special criticism. "True conservation practice would require the gas produced in the production of oil to be returned to the underground to restore pressure, or used commercially, or close in the wells. With such a program of true conservation, the amount of oil a given owner desired to bring to the surface and sell on the market should be left to the well owner's discretion."[10]

This conclusion is untenable: Correlative rights would be ignored, underground waste would result, and marginal wells would be prematurely abandoned. And it is incorrect to claim that all gas not returned to the reservoir or used commercially is wasted; it has performed the primary function of helping to lift the oil to the surface.

The limitation of oil production to market demand fosters conservation. Oil not yet produced but held back in flush fields able to produce efficiently at higher rates is actually as readily available to the market as oil in surface storage. And underground reserves are not subject to the physical and economic waste of surface storage subject to evaporation, deterioration, and the hazards of fire and leakage.

Governor Shivers explained to a Congressional committee in 1950 the reasons that prompted Texas to adopt a market-demand statute. He related how the market in 1932 was flooded with excess oil from the East Texas field, and stated:

> The Legislature of Texas met in special session and recognized that oil produced in excess of market demand was waste, and passed the market-demand statute, which said that production of oil in excess of transportation facilities or market demand was waste and instructed the Railroad Commission to limit production to that amount of oil that could be transported and marketed.

10 "Oil Supply and Distribution Problems," *loc. cit.*, 14.

In other words, the legislature recognized that it was perfectly foolish and wasteful to produce more oil than could be sold.

Immediately thereafter the Railroad Commission of Texas posted orders limiting the East Texas field, along with other fields of the state, to that amount of oil that could be produced without waste and marketed.

The court promptly upheld our order for the first time, stating that now the legislature had seen fit to authorize and empower the commission to limit production to market demand.

Since that date, every oil field in the State of Texas has come under the market-demand statute, and the market demand from the oil fields of Texas is ratably divided among the various fields of the state and then the amount which is allocated to an oil field is ratably prorated among the oil wells of that particular field.

This system of proration and waste prevention has proved highly successful, and has meant that by preserving the pressure in the reservoirs, hundreds of millions of barrels of additional oil will be recovered from the oil fields of Texas.

The East Texas field has already produced 2,600 million barrels and the pressures are holding up wonderfully.

When the field was first outlined by dry holes around its 130,000 producing acres, the experts rather uniformly predicted that the field would produce in the nature of two billion barrels of oil total ultimate yield. They were figuring on old methods.

Now it seems safe to predict that if the East Texas field is as carefully controlled in the future as it has been since 1942, the field should produce two billion barrels more of oil, and the grandchildren of young people living in East Texas today can have oil produced on their properties.[11]

In June, 1956, Ernest O. Thompson reviewed the method

[11] "Petroleum Study," *loc. cit.*, 725–26.

of ascertaining the market demand for Texas crude oil; similar procedures are followed by other states with market-demand statutes.

We ask for sworn offers from all purchasers of Texas crude —firm for the succeeding month, and best estimates for the next five months.

We receive from the Bureau of Mines their monthly estimate of the total amount of Texas crude required each day for the succeeding month. This oil can come either from production or from out of storage. There is no compulsion about the Bureau of Mines estimate. We can take it or leave it. We consider the estimate as evidence of market demand, along with the offers to buy certain amounts of Texas crude.

We then look to the report of the Bureau of Mines on the amount of Texas crude in storage above ground. If the amount of Texas crude in storage above ground is increasing excessively, then that is evidence that Texas is producing more crude than is being used. If stocks of Texas crude above ground are dropping appreciably, then that is evidence that Texas can produce more crude oil without causing physical waste in unnecessary storage above ground. This physical waste comes from evaporation, leakage, and fire hazards.

Oil keeps its lighter ends better in the reservoir than it does in above ground tanks. These lighter ends are the most valuable.

At the monthly statewide hearing all the above facts are put into testimony and in evidence. Any waste in any oil or gas field in Texas is a matter of inquiry and full discussion. Any person interested can ask questions about any matter in which he may have an interest.

Every word is recorded, and a transcript of the hearing is prepared. Anyone can purchase it.

When the facts are all in, the Commission decides upon the amount to be allowed to be produced each day for the succeeding month. For example, we decided that 3,400,000 barrels per day were needed to fill the demand for Texas crude for May, 1956.

We then first allocated 986,421 barrels to the statutory marginal and exempt wells, then allocated the balance to the various fields of the state according to ability to produce oil.[12] Then we prorated each field's allowable among the wells in that field according to the field allowable rule applicable to that field.

Each oil field has rules and regulations promulgated after notice and hearing on the producing characteristics of that particular producing horizon or horizons. These hearings are held on call of our Commission after five wells have been drilled. This gives the Commission actual field data upon which to prescribe rules based upon sound engineering principles.

. . . We have been to the court house many times, but with the passage of the market demand statute the Federal Courts have refused to interfere with these purely state administrative problems and processes in oil and gas conservation.[13]

Although state and federal courts have sustained state statutes relative to market demand, ratable takings, and physical waste, a few congressmen still view Article V of the Compact as a price-fixing device.[14] The Compact was renewed by Congress

---

[12] Stripper wells are exempt, as well as production from oil fields revived by water pressure. Of the latter, General Thompson stated before a Congressional hearing in 1953: "We limit them not at all because we figure that is bringing back the dead and is the highest type of conservation, reviving that which would go out on account of economic abandonment."

[13] Lieutenant General Ernest O. Thompson, "The Railroad Commission of Texas, Oil and Gas Conservation Authority—the Legal Basis of Its Operation," *The Oil and Gas Compact Bulletin*, Vol. XV, No. 1 (June, 1956), 8–10.

[14] Invariably, questions raised by some committee members at Congressional hearings on renewal of the Compact touch upon "price-fixing." See "Interstate Oil and Gas Compact," *loc. cit.* On page 107 of the report is a letter of June 15, 1951, from H. G. Morrison, assistant attorney general, in which he advised Hon. Robert Crossee, chairman of the Committee on Interstate and Foreign Commerce, that the Antitrust Division of the Department of Justice had found no violation of antitrust laws by member-states of the Compact. The Federal Power

in 1955 for a four-year period ending on September 1, 1959.[15] However, Section II of the Joint Resolution of July 28, 1955, concerning the Compact, recommended that the attorney general should make an annual report to Congress "as to whether or not the activities of the States under the provisions of such compact have been consistent with the purpose as set out in Article V of such compact." The first annual report was filed on September 1, 1956, by Herbert Brownell, Jr., U. S. attorney general, with only incidental reference to the Interstate Oil Compact Commission; it contains an inference, however, that the Department of Justice may conduct a more searching investigation on the competitive effects of the system by which oil production is regulated.[16]

The Compact Commission placed at the disposition of the attorney general all its reports, committee proceedings, and publications; and summary findings published in his second annual report paid tribute to the Compact Commission's "reputation and influence," and cited it as an "unique example of interstate cooperation on a wholly voluntary basis."[17]

Commission, the Federal Trade Commission, the Department of the Interior, the Department of Commerce, and the Secretary of Defense recommended its renewal. These federal agencies usually have personnel in attendance at Compact meetings.

[15] 69 U. S. Statutes, 385.

[16] *Report I of the Attorney General,* concerning the Interstate Oil and Gas Compact, September 1, 1956.

[17] *Report II of the Attorney General,* concerning the Interstate Oil and Gas Compact, September 1, 1957.

12

TECHNOLOGICAL ADVANCES

H. M. STALCUP, vice-president of the Skelly Oil Company, asserted in May, 1938, that "about 99.44 per cent of what has been accomplished along conservation lines has been done in the last decade."[1] He credited these advances to the co-operation of the industry with governmental agencies and to the work of petroleum engineers. He stated:

> Out of the circumstances of controlled production, aided and abetted by petroleum engineers, in the past ten years has come an almost endless chain of improvement in operating practices and mechanical perfection that seems almost incredible and conclusively confirms the belief that more progress has been made in the past ten years than in the balance of the life of the industry.
> Let us examine the record. Who, among you, ten years ago, had even so much as heard of the expression back-pressure, bottom-hole pressure, completion of wells through a one-eighth inch choke, producing one through a 1/64 inch choke, gas-oil ratio, fluid level, pressure maintenance, reservoir energy, controlled water drive, maintenance of uniform bottom-hole pressure, reservoir equilibrium; and so on and on with an almost endless list of terms now bandied about as though we had used them always, not just for the purpose of display-

[1] Stalcup, "What the Oil Industry is Doing about Conservation," *The Oil and Gas Journal,* Vol. XXXVII, No. 1 (May 19, 1938), 50.

ing an improvement of our vocabulary, but each and every one representing a vital step in the scientific progress in improving the technique of producing a higher percentage of the recoverable oil in a given pool, and at a lesser cost.

Similarly, in the mechanics of the job, ten years ago who of you had ever heard of straight hole drilling? We simply drilled a hole in the ground, letting nature take its course, whereas today, a deviation from the vertical in excess of 3 degrees is taboo. On the other hand who among us ever so much as dreamed a decade ago of intentionally controlled directional drilling? Today a well may, at will, be drilled with uncanny accuracy in any desired direction, so that the bottom of its hole may be at a predetermined point many hundreds of feet away laterally from the derrick floor.

Ten years ago we knew less than nothing of the importance of keeping the mud-laden fluid with which rotary drilling is done to any particular standard; whereas, today it is regarded as of the very essence of proper drilling and completing an oil well efficiently; and a roughneck member of a rotary drilling crew watches the weight and viscosity of its mud with the same intense interest as does a fond young mother the daily diet of her babe in arms.

How common ten years ago was the wire-line core barrel, an improvement in the art of cutting and bringing to the surface samples of the formation drilled through without which countless many valuable oil reservoirs would have remained undiscovered?

Who, much less than ten years ago, ever heard of an electrical logging device, handily run in a rotary drilled hole, still full of mud-laden fluid, to give the operator a picture of the permeability, porosity and possible oil, gas or water content of sands, dolomite or lime formations drilled through?

Or, who had heard of the practice of deliberately drilling through and casing and cementing off an oil sand, and then perforating such casing at a predetermined depth opposite such sand, with an electrically operated gun that shoots holes through the pipe at desired depths? This last mentioned method and development, now in use about five years, is truly a remarkable aid in conservation, since it makes possible the completion of oil wells with low gas-oil ratios in certain areas and under certain formational conditions well-nigh impossible of completion under the older and conventional methods.

And, who had heard of the now common practice of treating with hydrochloric acid, lime formations to increase the natural porosity and permeability, and thereby the oil and gas recovery? Truly remarkable results have been obtained by this method in very recent years, first, in revising the productive ability of old and therefore largely depleted lime wells, and second, by greatly increasing, often several hundred per cent, the natural flow of newly completed lime wells. Instances are on record where literally dry holes have been converted into profitable oil wells by this method, and countless noncommercial wells have been changed into profitable ones by such acid treatments. This would seem to be an instance of conservation with a capital "C."[2]

During the five-year period ending in 1939, 752 new oil pools were discovered in the United States. Only fourteen of the discoveries were made by random drilling; the others were the result of geological or geophysical exploration. When surface geology did not appear favorable, areas were reworked through geophysical checks. The torsion balance was introduced in 1922; the refraction seismograph was first used in 1924; and the gravity meter was introduced in 1931 as a means of ascertaining the unequal distribution of formations of different weights beneath the earth's surface. The seismograph is used to measure either refracted or reflected artificial sound waves induced through the detonation of high explosives in shallow holes, and seismograph crews have become specialists in oil exploratory opera-

[2] *Ibid.*, 50 ff.

tions. Sound waves travel at different velocities through different types of rock. On the Gulf Coast, for example, the velocity through clays, shales, and sandstone formations is normally about 6,000 feet a second, but through salt domes it is 15,000 to 16,000 feet a second—nearly fifteen times the velocity of sound in air. If a salt plug comes to within 5,000 to 6,000 feet of the surface, its presence is indicated by the arrival of sound waves refracted through the salt in a shorter time—sometimes as short as three-tenths of a second—than if a high-velocity formation had not been encountered.[3] In some instances, subsurface mapping has been completed to a depth of 25,000 feet. Specialization has included, too, the examination of soil air—the atmosphere near the ground—for traces of minute quantities of hydrocarbons and for a possible indication of oil-gas sand underneath. Because of advances in the knowledge of subsurface geology, oil was being produced by 1939 at depths of from 10,000 to 14,000 feet in twenty-five pools.

Deeper drilling has required the use of heavier and more expensive equipment. A complete drilling outfit—a 136-foot derrick, complete with draw-works, swivel, traveling block and crown block, rotary hose, drill pipe, two gas or diesel drilling engines, two mud pumps, mud tanks, shale shaker, blowout preventers, and light plant—may cost as much as $500,000. Truss wagons are sometimes used to skid the draw-works and derrick, which weigh as much as 800,000 pounds, from one drilling location to another.[4] In recent years, 178-foot derricks, with bases 32 feet square, capable of accommodating four miles or more of drill pipe, have come into use.

Drilling bits are now faced with harder metals and alloys to prolong their drilling life; a special grade of sintered tungsten carbide adds necessary qualities of toughness and abrasion to

[3] E. L. DeGolyer, "How Men Find Oil," *Fortune* (August, 1949), 97–103.

[4] An article by E. Lawson Lomax, "Expansion in Kuwait," in *World Petroleum*, Vol. XXVII, No. 11 (October, 1956), 55, relates how a heavy draw-works and 136-foot derrick were skidded over ninety-nine miles of desert.

the spherical-nosed compacts used as teeth. In deep drilling, from four to eight hours or more may be required to "come out of the hole" with the drill pipe, change bits, and "run back in"; hence the fewer times this is necessary, the less expensive the operation will be. When the Oklahoma City field was opened in 1929–30, the drilling time for a well with average footage of 6,500 feet was 70 to 90 days; recent extensions have been completed in less than 30 days. In 1951, it took 120 to 150 days in West Texas to drill wells 13,000 feet to production; now, they can be drilled in 70 days under similar conditions. Continuous studies are underway on the effects which the bit weight and rotary-table speed have on drilling.

Hines H. Baker, president of the Humble Oil and Refining Company of Houston, Texas, stated to a Congressional committee in 1953 that the average cost for each well completed by his company the previous year was $106,500, or 29 per cent more than the cost in 1948. W. W. Keeler, vice-president of the Phillips Petroleum Company of Bartlesville, Oklahoma, submitted evidence to the committee that drilling costs for each well completed by his company in 1952 were $114,800. Keeler pointed out the greater costs inherent in deeper drilling: "To increase the depth of a well from 3,000 feet to 4,000 feet, the additional 1,000 feet cost $11. per foot. To increase the depth from 7,000 to 8,000 feet, the additional 1,000 feet cost $20. per foot, and to increase the depth from 12,000 to 13,000 feet, the additional 1,000 feet cost $72. per foot. A 13,000-foot well, although only two and one-half times as deep costs over six times as much to drill as a 5,000-foot well."[5]

Savings in drilling and completion costs have recently been effected in production zones less than 3,500 feet deep by "slim-hole drilling." Lighter, portable rigs and equipment are used, and a smaller hole is bored by a smaller crew. Time is saved in setting up, tearing down, and moving to locations, casing costs

[5] "Petroleum Study," *loc. cit.,* 94, 158, and 276.

are reduced one-third, and other factors result in comparable cost savings.

Instruments and gauges have been developed which indicate the weight on the bit, the torque in the pipe, the speed of the rotary table, the pressure on the mud pumps, and the load on the derrick. A continuous record of the drilling performance is made. Drill-collar assemblies up to 360 feet in length are inserted in the drill stem to provide the weight necessary to prevent the drill from deviating from a vertical course and making a crooked hole.

The whipstock came into use to force the drill by pressure toward one side of the hole. The side pressure causes the drill to cut a hole that curves away from its original course. Directional drilling by this means has been effectively used to tap producing oil sands under industrial plants, schools, and other permanent buildings, such as the state capitol in Oklahoma City, and also offshore, as in the Wilmington–Long Beach area of California. Oil leases under the Pacific Ocean opposite Huntington Beach, California, have been developed from well sites several city blocks from the shore line, and multiple wells are being developed from single sites on barges in the Gulf of Mexico, off Louisiana and Texas.

Great improvements have been made in the mud-fluid control used in rotary drilling. Mud which is to be forced under heavy pressure down through the drill pipe to the bit to act as a cooling lubricant and carry the cuttings from the well has to have the proper characteristics. On its upward course, it moves between the pipe and the sides of the hole, and acts not only as a lubricant for the spinning pipe but also coats the side of the hole with a layer of mud that prevents loss of fluid from the formation and discourages enlargement of the well diameter. Chemists, or "mud engineers," test samples of the mud fluid for weight or filtration loss, viscosity, and gel strength, and see that the proper characteristics are maintained by chemical treatment. A mud mixture with these characteristics increases the life of drill bits, and thus reduces the number of times the pipe has to be pulled to change bits.

As drilling becomes deeper, the drilling-mud pressure increases—at a depth of 10,000 feet, a column of mud weighing 75 pounds a cubic foot exerts a pressure at its base of about 5,200 pounds per square inch—and there have been cases where the mud has sealed off oil-bearing sands.[6] The weight of the mud is usually sufficient to overcome pressures found at great depths, but sometimes the unexpected happens. In the La Belle field in Jefferson County, Texas, a well began producing through the drill stem, which could not be withdrawn when high pressure was met unexpectedly at 8,632 feet. Two months later the well was producing 937 barrels of oil daily and 79,000,000 cubic feet of gas—a ratio of 1 to 84,300. Six months later the well, still out of control, was producing 720 barrels of oil and 70,-000,000 cubic feet of gas, a ratio of 1 to 97,000.[7]

The unexpected has often happened in drilling for oil when gas, confined under enormous pressure, has been struck. Many stories are told of "wild" wells that have cratered and of the heroic efforts to bring them under control. In such instances, courts have ruled that the interest of the owner must yield to public necessity—just as a sheep-killing dog or a horse that has developed a fatal, contagious disease may be summarily destroyed.[8]

Recent technological imrovements have minimized the

[6] Levorson, *op. cit.*, 60.

[7] H. C. Miller and G. B. Shea, "Recent Progress in Petroleum Development and Production," 76 Cong., *House Resolutions 290 and 7372* (1939–40), 303.

[8] In *Davenport* v. *East Texas Refinery Company,* 127 S. W. (2d) 316 (April, 1939), the court held that 200,000 barrels of inflammable oil which had flowed into a creek in June, 1935, "loose and afloat upon flooded streams was [were] comparable to a mad dog loose in the community which called for action then and there." *Corzelius* v. *Railroad Commission et al.,* 182 S.W. (2d) 412 (July, 1944) describes in great detail the destructive force of a gas well out of control.

For a graphic description of work on a "wild" well, see John L. Flaherty, *Flowing Gold: The Romance of the Oil Industry,* 75–81.

danger of blowouts. These include pressure-tested casing which has been properly cemented, high-pressure fittings, dependable blowout preventers, quick-closing valves for flow line and drill pipe, large capacity, high-pressure slush pumps, adequate power, and an ample supply of properly prepared mud fluid in reserve tanks.[9]

Controlled-pressure drilling has been effectively used in the extension of fields where the bottom-hole pressure is known. Instead of a mud fluid, a mixture of oil and gas is used as the circulating medium at a pressure slightly lower than the reservoir pressure. This method has been used in the northern extension of the Oklahoma City field, in Kettleman Hills, California, in West Texas, and in New Mexico.

An electrically operated gun, to perforate casing at a predetermined depth, came into use in 1933. Another innovation has been the use of sections of magnesium-alloy pipe in casing off the upper oil-bearing sands. The pipe is soluble in hydrochloric acid, which can quickly eat away the pipe when the driller wants to expose the sand to production.

Advancements in engineering knowledge are utilized also by state regulatory commissions in assigning daily allowable production from newly developed fields. Most state agencies require either the filing of a permit or a sanction to drill. If the wildcat venture becomes a commercial producing property and the immediately surrounding acreage is controlled by different landowners or operators, the owner of the discovery well requests a hearing before the regulatory body to discuss field rules. At the hearing, technical testimony is given in regard to the geological formation, the characteristics of the producing sand and reservoir fluid, and the gas-oil ratio. Completion data of the well is studied, protection of fresh-water strata is stressed, a casing program is worked out, and consideration is given to the

[9] Miller and Shea, "Recent Progress in Petroleum Development and Production," *loc. cit.*, 302.

probable type and size of the field in relation to the diversification of land ownership. From these data, recommendations are made regarding proper well-spacing, drilling, and completion practices. The agency then issues an appropriate order, subject to revision when additional wells have been drilled.

Deeper drilling in recent years brought the discovery of a new type of petroleum accumulation known as "distillate" or "condensate" fields. A portion of the gas condenses in liquid form when the pressure is reduced, and unless reservoir pressure is maintained, a portion of the condensed liquid components are unrecoverable. Laboratory and field experiments and research have resulted in the development of a new producing technique known as "gas cycling." The reservoir fluid is produced in the gaseous phase at high pressure, liquid constituents recovered, and the dry gas compressed and returned to the producing formation, where it serves to displace the wet gas and to maintain reservoir pressure.

A factor in increasing the recovery of oil has been the unitization of oil fields. Henry L. Doherty, in a letter to President Calvin Coolidge on August 11, 1924, pointed out that, under the law of capture, each new pool or discovery had to be immediately devastated, and that, in most instances, a well produced more oil during its first year than in all succeeding years.

It took time, however, for unitization to be generally accepted, partly because of the fear that it was in conflict with antitrust legislation. In 1928, Amos L. Beaty, president of the Texas Company, proposed that unit operations be brought about by the organization of unincorporated trusts, and in 1930, the Mid-Continent Oil and Gas Association published its 141-page booklet, *Handbook On Unitization of Oil Pools.*

Ray Lyman Wilbur, secretary of the interior, in his annual report that year called attention to the waste attendant to the production of oil at the public reserve of Kettleman Hills: "Each year the waste of energy from that field is over twice the total expected annual electric output of Hoover Dam. Californians who for years have advocated the development of the Colorado River as a great new source of energy probably do not realize that failure to enforce strict conservation measures is costing them annually twice as much irreplaceable energy as Hoover Dam will be able to make good." Early in the following year, he approved a plan for the unit operation of the pool under the provisions of a temporary act of Congress. Not until August 21, 1935, did Congress enact legislation requiring development under a co-operative or unit plan for pools embracing public lands developed under leases.[10]

When the oil industry was depressed by overproduction in 1931, R. R. Penn, vice-president of the American Petroleum Institute and an independent Texas producer, offered a unique proposal: the formation of a giant corporation for the unit development and operation of the East Texas field similar to that formed to operate Kettleman Hills in California. His proposal was rejected, but engineers, geologists, and oil operators were in general agreement that the unit operation of several properties overlying a common source of oil or gas supply would increase recovery, reduce economic waste, preserve reservoir pressure, and prolong the life of the field. Any well which will not increase materially the total ultimate recovery from an oil or gas pool is an unnecessary well. The industry and state legislatures have accepted this fact; hence well-spacing and unitization statutes have been passed in oil-producing states.

In recent years, water-flooding has produced an ever increasing portion of the nation's oil. Secondary recovery by this method has been made possible by joint co-operation of the owners of production.

Corrosion of equipment is a major item in water-flooding projects; it has always been a serious problem in pipe-lining.

[10] 49 Stat., 676. See Robert E. Hardwicke, *Antitrust Laws et al.* v. *Unit Operation of Oil and Gas Pools,* for a fine exposition of this principle.

Recent experiments have proved that an installed cathodic protection system, using rectifiers and carbon anode ground beds, gives protection against external corrosion to a properly coated pipe line. Operating costs have been reduced, too, since 1950, by the limited use of plastic pipe where the pressure does not exceed 35 pounds per square inch. Plastic pipe is immune to electrolytic action, is noncorrosive, and remains free from paraffin. Experiments are underway to test the corrosion resistance of aluminum pipe, where pressure does not exceed 500 pounds per square inch, and without the usual coating, wrapping, and cathodic protection applied to steel pipe.

Pipe lines, of course, are preponderantly of steel construction. Until the late twenties, thread and coupling joints were used to join together twenty-foot lengths of pipe which were buried in trenches dug by manual labor. Then the industry adopted seamless and electrically welded pipe of higher tensile strength, thinner walls, and less weight per foot, and mechanical ditch-digging machines came into common use.[11]

In 1930, a group of Mid-Continent refiners—the Barnsdall Corporation, Continental Oil, Skelly, Pure Oil, Mid-Continent, and Phillips companies formed the Great Lakes Pipeline Company to transport gasoline to the Chicago marketing area.[12] The pipe line—1,350 miles long—was completed early the next year. At that time, the rate charged for gasoline hauled that distance by railroad tank cars was 2⅜ cents a gallon; the pipe-line rate was 1 to 1½ cents. By the late 1930's, it was the common practice of the industry to pump different batches of products through a line: as many as eight grades of gasoline, two grades of kerosene, two grades of furnace oil, and one grade of gas oil were being pumped consecutively through a line, with all the products delivered ready for use.[13] Pipe-line rates per barrel of these

[11] William G. Heltzel, *Pipeline Manual,* in Charles Morrow Wilson, *Oil Across The World,* 285.

[12] *Business Week* (January 7, 1931), 21.

[13] Wilson, *op. cit.,* 241. Crude oil gathering lines are usually two, three, or

products from Tulsa to Chicago are now 58 cents; by railroad tank car, $1.66. Crude oil pipe-line rates from Midland, Texas, to Houston are 18 to 21½ cents a barrel; to Chicago, 35 to 40 cents. Railroad tank-car shipments per barrel are $1.75 and $1.66, respectively. Ocean tankers haul crude oil from Houston to Philadelphia for 37½ cents a barrel; the railroad rate is $3.06.[14]

During the period since World War II, supertankers have been built to carry petroleum and petroleum products to meet the increased world demand. In 1937, only 21 per cent of all tonnage moved by ship was oil; now, 45 per cent is oil. Some tankers are already too wide to use the Panama Canal, too deep of draft for the Suez Canal when fully loaded, and so big that

four inches in diameter; tributary and trunk lines are six to twenty-four inches in diameter. William G. Heltzel, general manager of the Stanolind Pipeline Company, in the *Pipeline Manual, loc. cit.,* 284, states: "How much crude oil can be pumped through the sizes of pipe used in the main and tributary systems? Using a pressure of 12 psi per mile (which would be the equivalent of an 80-mile distance between stations, and a pumping pressure of 960 psi) for moving a 37-degree A.P.I. (gravity) 75 S.S.U. (Saybolt Second Universal) viscosity crude oil, the following capacities are determined by the turbulent flow formula using Stanton and Pannell data for 'f':

Nominal Diameter	Actual Inside Diameter	Capacity Barrels (42 gal. per bbl.) per 24 hours
6	6.065	7,730
8	7.981	16,800
10	10.02	31,200
12	12.00	50,400
14	13.25	66,000
16	15.25	97,800
18	17.438	140,400
20	19.374	188,400
22	21.250	245,000
24	23.250	310,000

"If more viscous crudes are used, these capacities would be materially reduced."

[14] Excerpts from a comprehensive table of rates that appeared in *The Oil and Gas Journal,* Vol. LIV, No. 46 (September 17, 1956), 140. Rates, of course, will vary according to industry and world economic conditions.

many have to discharge up to one-third of their cargo on lighters before entering most ports. When the *Spyros Niarchos,* a 47,750-dead-weight-ton oil carrier, was launched in England in December, 1955, it was the biggest tanker in the world; it broke all tanker records in July, 1956, by discharging a cargo of 41,000 tons of crude oil (287,000 barrels) at Rotterdam, in seventeen hours and forty-eight minutes. In March, 1956, the *Cities Service Baltimore* was launched; it is capable of carrying 273,175 barrels of eighteen different products. The 65,000-ton *Stavros Niarchos,* which is 850 feet long and has a 115-foot beam, can carry 550,-000 barrels of crude when fully loaded. The combination carrier *Sinclair Petrolore,* of 55,540 dead-weight tonnage, was launched at Kure, Japan, in July, 1955. Its crude-carrying capacity is 447,600 barrels, with additional storage space for 44,230 barrels of fuel. A self-unloading mechanism permits the discharge of 29,600 barrels of the cargo each hour. In August, 1956, the 84,730-ton *Universe Leader* was commissioned at Kure, Japan. At that time, more than 20 supertankers were under construction in the United States, England, Germany, Sweden, and Japan, each in excess of 40,000 tons. Two tankers, each of 87,-000 dead-weight tonnage, are now under construction in Japan. It is expected that by 1960 more than 350 supertankers, some in the 100,000-ton class, will be on the sea lanes. Tankers haul oil across the oceans for three to four cents a gallon, which is less than prewar costs. A 50,000-ton tanker carries few more crewmen than a 16,000-ton wartime tanker, is speedier, and is more efficient for every unit of horsepower. Supertankers are equipped with air-conditioning units; they afford comfortable quarters for the crew, as well as tiled showers, television, elevators, and recreation rooms.[15]

Offshore drilling in the Gulf of Mexico on the continental

[15] *Time,* Vol. LXVIII, No. 6 (August 6, 1956), 55–62; *The Oil and Gas Journal,* Vol. LIV, No. 46 (March 19, 1956), 132; Vol. LIV, No. 46 (September 17, 1956), 137; and Vol. LV, No. 2 (January 14, 1957), 90; *World Petroleum,* Vol. XXVII, No. 11 (October, 1956), 98.

shelf bordering the states of Louisiana and Texas has rapidly increased during the past decade. The shelf, an extension of the flat coastal plain, slopes very gradually from an elevation of about 600 feet above sea level at some interior points near the coast to about 600 feet below sea level, and extends out into the gulf 70 miles from the mouth of the Río Grande and 140 miles from the mouth of the Sabine River. David White, of the United States Geological Survey, expressed the opinion as early as 1927 that anticlines and salt domes would be discovered when exploration was made of the shelf beneath the Gulf, and in 1948, E. L. DeGolyer, the well-known geologist, estimated that if domes were as common offshore as onshore, there might be 10,500,000,000 barrels of reserves off the Louisiana–Texas coast.[16]

Dean McGee, president of Kerr-McGee Oil Industries, Inc., of Oklahoma City, in 1954 stated the reasons for the postwar interest in exploratory operations in the area:

> Since the discovery of oil at Spindletop in 1901 more than five hundred fifty-six producing structures have been found on the land portion of this shelf in an area of approximately thirty-nine thousand square miles extending westward from the Mississippi River to the Rio Grande and extending inland from the coast from twenty to ninety miles. An equivalent area in the Gulf would extend from the shoreline out to a water depth of about one hundred sixty feet. It would have a maximum width of ninety-eight miles off the Sabine River.
>
> The development of these five hundred and fifty-six structures has required the drilling of about thirty-six thousand development wells and resulted in the discovery of something over eleven billion barrels of oil and sixty-seven trillion cubic feet of gas. Approximately thirty-seven thousand dry holes have been drilled. Of these five hundred fifty-six oil fields, sixteen will have in excess of one hundred million barrels recov-

[16] "Investigations of Petroleum Resources," *Senate Resolution 36, loc. cit.;* "The Big Gulf Gamble," *Fortune,* Vol. XXXVIII, No. 1 (July, 1948), 69–72.

ery, twenty-nine will have recoveries between fifty and one hundred million barrels, and fifty-three in the twenty-five to fifty million bracket, or a total of ninety-eight fields that will have an ultimate recovery in excess of twenty-five million barrels.

The oil is being produced from three principal types of structures—salt domes, faulted structures not attributable to salt movement, and lens or stratigraphic types. All available data indicate that the same type accumulations will be found in the immediately adjacent continental shelf area.

While available geological information does not permit an accurate estimate of the distance offshore that favorable stratigraphic and structural conditions exist, offshore well data indicate that nearly all formational units are continuing to thicken seaward. Geophysical exploration has disclosed favorable structural conditions far at sea. Certainly, exploration to date has shown the offshore area to be comparable, in number and quality of salt dome structures, with the coastal belt.

While the gravitymeter, airborne magnetometer, core drill and refraction seismograph have been used to some extent, substantially all offshore exploration has been and is being done with the reflection seismograph. Water seismic techniques and equipment are very efficient and now include tow cables, radio surveying, packaged explosives and the use of large boats. . . . As instruments, techniques and skill in interpretation have improved through the years since 1930, the finding power of the reflection seismic method have increased many fold. Many coastal land areas have been surveyed again and again with the seismograph following each such significant improvement. Operators in the tidelands have available all the tools and techniques, as well as the experience accumulated, to apply to the search for the same type structural and stratigraphic accumulations.[17]

[17] Dean A. McGee, "Economics of Offshore Drilling in the Gulf of Mexico," paper for presentation to a group session of the Division of Production during the 34th annual meeting of the American Petroleum Institute in Chicago, on November 9, 1954. (Mimeograph copy, A. P. I.)

221

The first offshore drilling in this country was undertaken at Summerland, California, in 1896, from piers. In 1927 a drilling crew at Huntington Beach drilled a slanted hole seaward; after this, directional drilling became a common practice. At first the rotary table was simply tilted; then, in 1933, whipstock tools were invented for this purpose. Along the Gulf Coast, the first offshore drilling occurred in the 1920's from "islands" built in the shallow water by dredges; in 1938, the Superior Oil Company and the Pure Oil Company jointly built a platform in fourteen feet of water off the coast of Louisiana and struck oil 6,200 feet down from a salt dome structure.[18] This discovery of the Creole field, about one mile offshore in the open Gulf, led to the construction of platforms from which several wells were directionally drilled from a single location. In 1941, the Sabine Pass field, about 9,000 feet off the Texas shore, was developed. This was followed by developments in Galveston Bay and Laguna Madre, waters which were claimed by Texas. The question of tidelands ownership became a national issue, and ultimately a political football: Did offshore waters and the resources of the soil thereunder belong to the federal government or to the adjoining states? Millions of dollars of revenue from leases and royalties were at stake. President Harry S. Truman issued a proclamation on September 28, 1945, setting forth the policy of the United States with respect to natural resources of the subsoil and the sea bed of the continental shelf. He laid claim for the United States to the area and in later informal addresses stated that any revenue therefrom should accrue to all the states for education and other purposes. He proposed that revenues from the states claiming jurisdiction over the tidelands be divided in accordance with provisions of the Minerals Leasing Act: 37.5 per cent to the adjoining state and the rest to the federal treasury.

Immediately, the affected states took issue with the point

18 "The Big Gulf Gamble," *loc. cit.,* 71.

of view of the administration, and states' rights advocates rallied to their defense. Louisiana enlarged her boundary to a line twenty-seven miles from shore line; Texas claimed historic precedent for its boundary, three leagues or ten and one-half miles seaward. Nevertheless, the United States Supreme Court decided on June 23, 1947, that California could not claim rights to the continental shelf beyond the three-mile limit; similar decisions affecting Louisiana and Texas were made on June 5, 1950.[19]

The majority opinion in the Texas case pointed out that when Texas was annexed in 1845 the resolution provided it should be on equal footing with other states, that this negated any implied, special limitation of paramount powers of the United States, and that "Texas relinquished all claims to the marginal sea, including oil thereunder, to the United States." Justices Reed and Frankfurter prepared minority opinions, and the dissent of the latter included this statement: "As is made clear in the opinion of Mr. Justice Reed, the submerged lands now in controversy were part of the domain of Texas when she was on her own. The Court now decides that when Texas entered the Union, she lost what she had and the United States acquired it. How that shift came to pass remains for me a puzzle."[20]

The puzzle of tidelands ownership was not resolved by the court decisions. Twice Congress passed legislation to extend state boundaries offshore; in each instance, President Truman vetoed the measures. The question became an issue in the presidential campaign of 1952, and the Republican party platform provided: "We favor restoration to the States of their rights to all lands and resources beneath navigable inland and offshore waters within their historic boundaries." At Long Beach, California, on October 9, at New Orleans, Louisiana, on October 13, and at

[19] *United States* v. *State of California,* 332 U. S. 19 (June, 1947); *United States* v. *State of Louisiana,* 339 U. S. 699 (June, 1950); *United States* v. *State of Texas,* 339 U. S. 707 (June, 1950).

[20] *United States* v. *State of Texas,* 339 U. S. 724 (June, 1950).

Lubbock, Texas, on the following day, Dwight D. Eisenhower stated: "Twice by substantial majorities both Houses of Congress have voted to recognize the traditional concept of State ownership of these submerged areas. Twice these acts have been vetoed by the President. I will approve such acts of Congress." Four months after his inauguration, President Eisenhower approved the Congressional act which established title of the respective states to lands and natural resources three miles offshore.[21] Still to be decided by either Congress or the Supreme Court is a true determination of the Gulf shore line and recognition of the historic boundaries for Louisiana or Texas.[22]

Oil companies through 1955 had invested more than $1,250,-000,000 in all phases of offshore development. Prices for leases have pyramided: early in 1956, the Humble Oil and Refining Company bid $7,660,111 for a 4,687-acre lease in Louisiana, and the Magnolia Petroleum Corporation paid Texas $2,209 an acre for 1,440 acres off of the Texas coast. Drilling costs, too, because of the location and depth, are unusually high. The Shell Oil Company and the Continental Oil Company jointly drilled a 13,386-foot dry hole in 93 feet of water, sixty-three miles southeast of Galveston, which cost approximately $1,000,000. The Phillips Petroleum Company, in March, 1956, brought in a well drilled in 85 feet of water, forty miles off of the Louisiana coast. The well struck production of 1,300 barrels of distillate a day at 14,808 feet.[23]

Different methods of drilling are used at sea. Barges with living quarters are successfully used in shallow waters and are anchored there by pilings. A mobile platform or submersible

---

[21] "Act of May 22, 1953," 83 Cong., 1 sess., *Public Law 31,* 67 United States Statutes, 29.

[22] Texas expects a campaign against its claim to the historic boundary of ten and one-half miles because a decision by its Supreme Court set the common-law boundary at three miles.

[23] *The Oil and Gas Journal,* Vol. LIV, No. 46 (March 19, 1956), 131; Vol. LIV, No. 72 (September 17, 1956), 131.

unit which can be moved from one location to another is used in wildcat operations. Self-contained platforms, with storage space, from which may be drilled directionally multiple wells, are used after production is found. A recent trend has been toward the use of a combination drilling platform and drilling tender. A Kerr-McGee tender recently placed in operation ("Kermac Drill Tender III") has storage facilities for 13,000 barrels of drilling water, 24,000 gallons of Diesel fuel, two Emsco D-1,000 mud pumps, a Halliburton cementing unit, pipe, cement, and bulk drilling mud, as well as living quarters and a heliport. The main power plant, consisting of two diesel engines and generators, and an auxiliary plant are on the tender. Drilling equipment on the platform includes two 700-horsepower electric motors, the derrick substructure, and a 30x30x140-foot rig of 1,100,000-pound capacity. Multiple wells can be drilled directionally.

Like the modern ocean tankers which provide comfortable living quarters for their crewmen, most of the tenders now employed in offshore drilling have fully air-conditioned mahogany-paneled quarters for forty-eight men. They have one or more recreation rooms, large cold-storage rooms, a galley, and a mess hall. Some include a four-man hospital and a helicopter deck located over the quarters.[24] Every precautionary method is used to protect crewmen from accident or fire hazards. A stand-by boat is near the rig at all times for emergency evacuation; FM radio and marine telephone facilities provide weather news and communications; some operators even equip their personnel and work boats and platforms with radar.[25]

The greatest advances in the oil industry have been made in the last decade through communication and automation. The National Petroleum Council announced recently that wire and

[24] James W. Calvert, "New Floating Tenders for Offshore Work," *World Petroleum, loc. cit.,* 66–67.

[25] McGee, *loc. cit.;* also, see Alfred M. Leeston, *Magic Oil: Servant of the World,* 36–37.

radio facilities of the oil and gas industry could be linked into a nationwide network in case of a national emergency. There are estimated to be 41,000 transmitters in use by the industry, with $50,000,000 invested in microwave systems. The petroleum industry owns more than 170,000 miles of wire lines and has 55,000 miles of line under lease. Fred A. Seaton, secretary of the interior, has approved the proposal that these facilities be set up as a national network in the event of an emergency.[26]

Many pipe lines now use a microwave communication system, "a telephone system without wires." Although the pipeline walker, equipped with boots or snowshoes, who walks a line looking for signs of leakage, has not disappeared—in the Southwest, where more than three hundred flying days each year are common, the patrol can be maintained by air—automation is assuming complete control of the flow of petroleum and petroleum products through the lines. Metered volumes can be transmitted to remote counters or recorders; an operator at a distant point can gauge tanks and operate valves. Sound waves can be bounced off the surface of the liquid in a storage tank and, through a completely automatic ultrasonic system, the time required for the pulse to travel the path to the liquid and back is measured and converted to represent the liquid level to within one-hundredth of a foot. Pumping and compressor stations can be electronically controlled, and automatic samplers can report on the products handled.

Intricate control instrumentation is required in modern refineries and petrochemical plants to control temperature, pressure, and flow of liquids and gases; and electronic indicators, recorders, and controllers regulate the conversion of petroleum into by-products. The oil industry, too, has benefited from the advances made in all-purpose digital computers, which can make engineering and scientific computations where rapid calculation is important, and then can store on magnetic tape units

[26] *The Oil and Gas Journal,* Vol. LIV, No. 76 (October 15, 1956), 96.

the information that must be readily available. Data collected by seismographic crews, for example, can be assembled, stored at a central point, and made readily available as guidance for other crews.

The authors of a recent book have asserted that oil and gas production reached maturity with the invention and application of the bottom-hole-pressure measuring tool, which permits forecasts of pressure changes and estimates of the total recovery of oil and gas from a producing zone.[27] Daily improvements in techniques are made; any periodical devoted to the trade shows this trend. For an industry that has contributed so much in ancillary needs—from the manufacturing of tiny washers or microscopic scientific parts to a complete refining unit—it is not unexpected that advances in tools and equipment are constantly being made. There have been recent advances, for example, in combating evaporative loss from aboveground storage: plastic-treated coatings inside spherical-shaped storage tanks have reduced the incidence of evaporation. An advertisement of a product by another company noted that 1,200 barrels of crude oil were saved in a single year by a one-inch-thick blanket made of millions of microscopic phenolic plastic spheres filled with inert nitrogen gas, which floated on top of the oil stored in a 35x115-foot tank. Another company recently placed on the market tanks made of rubberized nylon, to be used for the storage of water, crude oil, and refined oil during fracture-treating operations in areas where water is not readily available. The tanks are 45 feet long, 11 feet wide, and 5½ feet high; they weigh 1,350 pounds and have a capacity of 15,000 gallons. When empty, they can be rolled up for easy transportation. Another new and convenient way to store up to 50,000 gallons of water above the ground has been developed. A circular metal fence about 4 feet high can easily be transported and put in place.

27 *Oil and Gas Production,* compiled by the Engineering Committee of the Interstate Oil Compact Commission, 82.

This is then lined with a covering of expendable polyethylene film, three-one-thousandths of an inch thick, which makes a watertight container ready for use in less than three hours' time.

Many of the technological advances are being made in the interest of conservation: for better recovery of oil from underground and for better utilization of the produced product. In 1955, the United States consumed oil at a rate five times the rate of discovery of new fields in the country. Because of the dependence on oil and gas as energy sources, attention has recently centered on sources of supply outside the continental United States.

13

THE PROBLEM OF IMPORTS

RUSSIAN OIL from the Baku field on the Caspian Sea and the Maikop District on the north side of the Black Sea near Crimea began to compete with American oil on the European market in the 1890's. This decade marked the beginning of an active interest on the part of American oil companies in the development of foreign oil reserves—first, before 1900, in Mexico, where flush production was found in 1910; then in Venezuela three years later, followed by the opening of the great Maracaibo Basin in 1917. American interests during this decade expanded into the Dutch East Indies, entered Iraq in 1927, and made important discoveries in Saudi Arabia, in Kuwait, and on Bahrein Island between 1936 and 1939.[1]

Oil in limited amounts has been imported into the United States since the first Mexican oil was shipped in 1910. This, as well as the greater percentage of all oil imported, has been owned by American companies. The principal quantities have come from Venezuela, which is closer to the greater refining centers in the Philadelphia–New York area than is Houston, Texas.

Not until 1930, however, when the states began seriously to limit domestic production, was much concern expressed over the importation of crude oil. During the last months of that year, as prorationists and antiprorationists marshaled their forces, the problem of imports became interrelated with the problem of

[1] Max W. Ball, *This Fascinating Oil Business*, 300–79.

proration. Events which transpired in the Mid-Continent area at that time reflect the interrelationship.

On December 4, 1930, the second anniversary of the Oklahoma City field, which at that time extended seven miles north and south and was two and one-half miles wide, the 653 producing wells were on proration at 95,000 barrels daily. Their potential production was listed as high as 5,000,000 barrels daily, which was much more than enough to supply the entire demand of the United States. During the week, a well came in at the northwestern edge of the pool that would gauge 65,000 barrels in twenty-four hours. Three days later a producer was struck on the Wheeler school ground, that was capable of producing 74,000 barrels in the same span of time; shortly thereafter, one capable of producing 90,000 barrels daily was struck.

On December 5, the Prairie Pipeline Company, operator of the most extensive pipe-line system in the world, with 12,500 miles of gathering and trunk lines extending through the Mid-Continent area, announced that, beginning on the first of the year, it was withdrawing from the market. This would affect 11,000 wells in Kansas, 20,000 wells in Oklahoma, and 1,900 wells in the Texas Panhandle.

Operators, restive under the 2.97 per cent allowable invoked by the Corporation Commission on production from the Oklahoma City field, awaited with unease a December 17 federal District Court decision in an injunction suit filed by an integrated company. The company could rightfully assert that there would be no waste from producing its wells in excess of the allowable because its pipe lines would carry the oil to its refinery. Another company, also, sought an injunction against the state's oil proration act. On December 8, a company opened a well for one hour and produced 1,200 barrels; another ran 6,431 barrels in two hours and thirty-five minutes. The Corporation Commission was considering placing all of the offending properties under receivership.

On Sunday, December 14, a full-page advertisement appeared in the Oklahoma City *Daily Oklahoman,* dealing with the "evils of proration." This was rebutted on Friday, December 19, with a full-page advertisement in the same newspaper by the Mid-Continent Oil and Gas Association, which, in summary, stated: "Proration is the only protection of the small producers who have no pipelines, refineries, or exclusive markets of their own." And this argument was supplemented the following Sunday with a full-page advertisement on the theme, "What will happen if proration is abandoned?"

The Mid-Continent Oil and Gas Association, in annual meeting at Dallas, on December 6, approved a resolution for a protective tariff on oil. On Sunday, December 7, E. B. Reeser, president of the American Petroleum Institute, joined the proponents of a tariff.

Representative Milton C. Garber, an Oklahoma congressman, three days later attacked proration on the floor of Congress. He declared that "proration puts Oklahoma in the national spotlight and in a more ridiculous position than the biennial impeachment of her governors." He characterized the procedure as "the most flagrant abuse and misuse of power in evidence throughout the nation." "Proration for conservation," he said, "is the propaganda of the Big Three—Royal Dutch Shell, the Gulf, and Standard. Proration, in fact, means our home markets for the cheaply produced foreign oils coming in duty-free." The congressman quoted figures to show that oil could be produced in Venezuela for 18 cents a barrel, delivered to a terminal for 22 cents, and shipped to the East Coast for 35 cents—a total charge of 75 cents a barrel—whereas the cost of producing and transporting one barrel of Mid-Continent oil to the same destination amounted to $1.75.

On December 9, Representatives W. H. Sproul of Kansas and Wilburn Cartwright of Oklahoma introduced a resolution in Congress to authorize a flat duty of $1.25 a barrel on imported

231

crude oil, 50 cents a barrel on fuel oil, and a 50 per cent ad valorem duty on refined products. Evidence was submitted that the importation of oil had increased from 58,000,000 barrels in 1927 to 110,000,000 barrels in 1929, that the United States furnished 70 per cent of the world's market demand for oil, and that its production was more than ample to meet the nation's needs.

Thus, in an atmosphere of bewilderment, as the great depression fastened its hold upon the nation's industries, with overproduction of crude oil and lessened use of its by-products, with the first stern testing of proration laws amidst conflicting court decisions, with many independent operators witnessing a growing tendency on the part of the big integrated companies to use more and more of their own oil and less of others', with the principal purchaser of the Mid-Continent area, the Prairie Pipeline Company, curtailing buying activity because its storage tanks were full and Standard no longer dependent upon it as a source for its Whiting, Indiana, refinery, the oil industry and, particularly, the independent producers looked to Congress to provide some tariff protection as a stopgap to their troubles. Congress, however, failed to enact a tariff measure.

Members of the Independent Petroleum Association of America and the Mid-Continent Oil and Gas Association were particularly active in seeking an import duty on oil. Memberships in these bodies were largely confined to the Mid-Continent and Gulf Coast areas which in the early 1930's accounted for more than 70 per cent of the nation's production. Independent operators controlled most of the Oklahoma City field, the Kansas industry, and virtually all the stripper wells in these states and in the Texas Panhandle, while major companies held most of the production in the Greater Seminole area in Oklahoma, in West Texas, and in New Mexico, and were gaining control of the giant East Texas area.[2]

[2] Oklahoma City *Daily Oklahoman,* August 2, 1931, Section B, 6.

Many of the independents are small producers, owners of stripper wells, who, by the nature of the industry, have no voice in the setting of prices, the sale of their oil, or the final distribution of its products. A few own refineries of small capacity, but are largely dependent upon major companies for their continued existence. Their philosophy was ably summed up in testimony before a Congressional committee: "We might ask you if that is the idea, that small business shall not be destroyed. There is no reason to make a featherbed for small business. We think this is a country for competition and hard competition, but we don't think it is exactly a jungle. We don't want to be devoured."[3]

Competitive advantages of major integrated companies have been summarized as follows:

> Cheap means of transportation, by pipeline and by tanker; ready access to supplies of crude oil, assured by ownership of crude and by access through crude oil trunk lines; assured marketing outlets from a highly integrated marketing program, which may involve exclusive dealer contracts and the use of short-term cancellation clauses in leasing and subleasing arrangements; tire, battery, and accessory programs which provide a profitable source of income without requiring proportionate investment; and the opportunity to diversify risks, to realize income from various branches of the industry's activities, including production, transportation, refining, and marketing.[4]

Although the cleavage between the independent and the integrated company is an ancient one, coexistent with the industry, the great depression forced governmental intervention and the use of untried measures to avert disaster. In the Mid-Continent

[3] Hearings before the Special Committee to Study Problems of American Small Business, Field Hearings in Kansas City on September 26, 1947, 80 Cong., *Senate Resolution 20* (1948), Part 20, p. 2,193.

[4] Statement of John Carson, commissioner of the Federal Trade Commission, May 19, 1950, in "Petroleum Study," *loc. cit.*, 261.

area, where independents favored proration—the Mid-Continent Oil and Gas Association passed a resolution at the Dallas meeting that production from any pool should not exceed market demand—the feeling persisted that the elimination of oil imports, which amounted to about 10 per cent of domestic production, would revive the domestic market. Some of the independents felt, however, that domestic conservation policies had been promoted by Standard of Indiana, Standard of New Jersey, Royal Dutch Shell, and the Gulf companies in order to have greater outlets for cheap Venezuelan oil.[5] In California, however, Standard Oil of California applied an old remedy to production fever: when operators in the Santa Fé Springs field failed to agree to an oil conservation program accepted by operators in other fields of the state, the company, in control of the market, announced reductions of from 75 to 90 cents a barrel in that district.

Governor William H. Murray of Oklahoma issued an invitation to oil-state representatives to meet in Fort Worth on February 28 and form the Oil States Advisory Committee. Representatives from the states of Arkansas, California, Kansas, Louisiana, Oklahoma, Texas, and Wyoming and from the oil industry met at Texarkana, Texas, on March 9 to discuss problems relative to the overproduction and importation of oil. The clash of interests between domestic proration and foreign flush production was clearly evident. Representatives of the Carter Oil Company, a subsidiary of Standard Oil of New Jersey, asserted import restrictions must be a permanent accessory to any coordinated proration plans, while representatives of the Pan-American Petroleum and Transportation Company, a subsidiary of Standard Oil of Indiana, stated that any proposed restrictions on imports should continue only so long as domestic output was held down.

[5] "Many Industries Have a Stake in Oil's Proration Battle," *Business Week* (January 28, 1931), 5–6; Oklahoma City *Daily Oklahoman*, December 8, 1930.

234

As a result of ideas formulated at this meeting and at the meetings later held by representatives of the Oil States Advisory Committee with President Herbert Hoover and officials of the federal Oil Conservation Board, the administration used its influence with oil-importing companies for a voluntary reduction of 25 per cent in the amount of oil brought into the country.[6]

Despite the voluntary curtailment of imports, domestic demand was exceeded by domestic supply, and, as aboveground storage accumulated, prices declined. In July, 1930, Kansas and Oklahoma produced 21,513,000 barrels of oil, worth $27,752,000 at the wells, or an average of $1.29 a barrel. In July, 1931, the two states produced 19,443,000 barrels of oil worth, at the wells, only $5,319,000, or an average price of 27 cents a barrel. Both states depended upon a gross production tax on oil to help support the needs of government. In July, 1930, this revenue amounted to $832,000; in July, 1931, to $160,000.

Actually, the two states were fairly successful in controlling production within their borders, but their programs were seriously affected by greatly increased production in Texas. "Dad" Joiner, after repeated failures, brought in the discovery well in the East Texas field on October 3, 1930; by mid-summer its daily expansion was proving it the giant of all discoveries, and the efforts of the Railroad Commission to control production, under Texas conservation laws, were unavailing. On July 28, 1931, at Houston, an opinion was handed down in federal court by Circuit Judge Joseph C. Hutcheson, Jr. and Judges Du Val West and Randolph Bryant, declaring invalid the April orders of the Commission applicable to wells in the East Texas field. The decision was a marked rebuke to the regulatory agency:

> Certainly when a subordinate body like the Railroad Commission of Texas undertakes as here to deal in a broadly restrictive way with the right of a citizen to produce the oil

[6] Murphy (editor), *op. cit.*, 549.

235

which under the laws of this state he owns, it must be prepared to answer his imperious query, "Is it not lawful for me to do what I will with mine own" by pointing to a clear delegation of legislative power. This must be found not in the recitative portions of its orders, for the commission may not more than any other agent, lifting itself by its bootstraps, supply, by claiming the power it does not have, but in the statutes themselves, which have created, which control, and which are the source of the commission's power.

Especially must this be so when, as here, under the thinly veiled pretense of going about to prevent physical waste the commission has, in cooperation with persons interested in raising and maintaining prices of oil and its refined products, set on foot a plan which, seated in a desire to bring supply within the compass of demand, derives its impulse and springs from and finds its scope and its extent in the attempt to control the delicate adjustment of market supply and demand, in order to bring and keep oil prices up.

We have searched but we can not read in any legislative pronouncement support for what the commission has done here.[7]

One week later at Guthrie, Oklahoma, a federal court held a diametrically opposite view in regard to proration orders of the Oklahoma Corporation Commission. Circuit Judges Orie L. Phillips and John H. Cotteral, with Judge Franklin E. Kennamer in dissent, found the oil proration law and orders issued by the Commission valid, although exception was taken to the penalty provisions of the orders.[8] It was the opinion of the court that "the limiting of takings to the market demands is a reasonable requirement for the prevention of waste and the protection

[7] *MacMillan et al.* v. *Railroad Commission of Texas et al.,* 51 Fed. (2d) 400–405 (July, 1931).

[8] Oklahoma City *Daily Oklahoman,* August 6, 1931. On May 16 of the following year, the United States Supreme Court affirmed this decision. See *Champlin Refining Company* v. *Corporation Commission of Oklahoma et al.,* 286 U. S. 210 (May, 1932).

of co-equal rights of the owners of lands over such pools."

Governor Ross Sterling of Texas had called a special session of the legislature in July to consider revision of the oil-conservancy law, and two measures were under consideration for regulation of production on the basis of market demand when the federal court decision was made at Houston. This decision undoubtedly influenced the Governor and the legislature, and the statute on conservation, passed by the legislature on August 12, specifically denied the limitation of production to market demand.[9]

Governor William H. Murray of Oklahoma, who dramatized his office, enjoyed the front-page coverage of his reference to the "inferior" federal courts—those other than the Supreme Court. When informed that there was a plot among certain legislators to prefer impeachment charges against him, Murray made headlines with the statement that that would be like a "bunch of Jackrabbits jumping on a wildcat," and issued an ultimatum in July to the major purchasers of oil in the state to raise the price of crude. He hinted that, unless this was done, he would issue a shutdown order to cut off the supply.

This was done on August 4. That night and the next morning the Oklahoma National Guard set up "off limits" areas in the principal fields of the state, in keeping with the executive order.[10] The state of Kansas, which earlier in the year had

---

[9] Robert E. Hardwicke, "Legal History of Conservation of Oil in Texas," in *Legal History* of *Conservation* of *Oil* and *Gas*, 229–33, 237. Hardwicke, a Fort Worth attorney, appeared before a committee of the Texas Senate on July 31 and made a strong plea for the inclusion of a market-demand provision in the proposed measure, similar to the provisions of the Oklahoma statute. The Texas Senate, by a vote of 16 to 9 on August 8, killed this provision. At this time, production in the East Texas field was approximately 1,000,000 barrels daily.

[10] The executive order included unnecessary allegations to the effect that a major oil company, during the previous March, while the legislature was in session, had invited forty legislators to a meeting in Tulsa to formulate plans for the governor's impeachment. It cited constitutional provisions in regard to school lands (in certain instances leased to oil companies, the revenue from which went to a school fund), and statutory provisions relating to gross production taxes and to waste, and specifically cited Chapter 25, Section 2, of the Session Laws of 1915: "That the taking of crude oil or petroleum from any oil-bearing sand or

adopted a conservancy measure copied largely from the Oklahoma statute, shut down the prolific Ritz-Canton pool and reduced ratable takings in other fields. Thirty-seven of the sixty principal operators in the East Texas field met at Tyler, Texas, on August 6 and voted in favor of a shutdown, to begin on August 14. Governor Sterling, who at first announced he was opposed to the recommendation, declared martial law for the field on August 17, when the price of oil there declined to ten cents or less a barrel. The Railroad Commission issued an order on September 2 for the East Texas field based upon the recently revised conservation statute to limit production to 400,000 barrels daily, and the National Guard remained in the field to enforce the order until a federal court decision on February 18, 1932, ruled that the governor did not have the right to control production by force of arms.[11] Not until Governor Sterling called a special session of the legislature, which passed the Market Demand Act on November 12, 1932, and included in its specific definitions of waste "the production of crude petroleum oil in excess of transportation or market facilities, or reasonable market demand," was proper authority placed with the Railroad Commission to regulate proration.

Oklahoma, in the meantime, was winning a series of court battles in regard to proration. The single important exception made it necessary to set ratable takings from each of the producing horizons in the Oklahoma City field rather than using the former method of treating the field as a common reservoir.[12] This field had remained shut down and under martial law from August 5 to October 10, and during the next two years the Gov-

sands in the State of Oklahoma at a time when there is not a market demand therefor at the well at a price equivalent to the actual value of such crude oil or petroleum is hereby prohibited. . . ."

[11] *E. Constantin et al.* v. *Lon Smith et al.,* 57 Fed. (2d) 227 (February, 1932).

[12] *H. F. Wilcox Oil and Gas Company* v. *State et al.,* 162 Oklahoma Reports 89 (February, 1933).

ernor used the National Guard to enforce the orders of the Corporation Commission or to ferret out the production of excess or "hot oil." Kansas, meantime, was effectively administering its proration statute.

Concurrently with attempts to control production in the flush fields of the Mid-Continent area and the regulation of production in California under the oil-gas-ratio law, the Independent Petroleum Association renewed its fight to obtain a tariff on oil. Recommendations were embodied in the Internal Revenue Act, passed by Congress on June 6, 1932.[13] This provided for a tax of 21 cents a barrel on petroleum and all liquid derivatives with the exception of gasoline, which was set at $1.05 a barrel, and lubricating oil, which was $1.68 a barrel.

The Petroleum Code established under the National Industrial Recovery Act of 1933 authorized a quota system on oil imports, and on September 2, Harold L. Ickes, secretary of the interior, announced that the importation of petroleum and petroleum products was not to exceed the average daily imports during the last six months of 1932. This amounted to 98,000 barrels daily, or 4.5 per cent of the domestic requirements.[14]

However, the quota system or importation problem was submerged under the more serious domestic supply system of illegal or "hot oil." The states, even with conservation officers and militia, were unable to cope with the maze of secret pipe lines, the stealthy runs to stills, and the entrance into interstate commerce of illegally produced petroleum and petroleum products. Federal legislation in 1933 which brought actual control in California by federal authorities, a voluntary compliance committee in Michigan, and control through other states' agencies curbed illegal production so long as the NIRA was in effect. On February 22, 1935, Congress passed the Connally Act, which was

[13] 47 United States Statutes at Large, 169.
[14] Northcutt Ely, "The Government in the Exercise of the Power over Foreign Commerce," in Murphy (editor), op. cit., 657.

scheduled to expire on June 16, 1937, but which was later extended with no stipulated ending date. Its administration solved the "hot oil" problem.

By the late 1930's the pattern of domestic crude oil production control by state regulatory agencies had been clearly established through administrative orders and court decrees. Little opposition was voiced by the industry to the trade agreement made with Venezuela in 1939, which cut in half the import duty on crude oil so long as the amount imported did not exceed 5 per cent of the crude run to domestic refineries during the preceding calendar year. The agreement provided that if the quota admitted exceeded that amount, the statutory tariff of 21 cents a barrel should be paid.[15] In 1938, the United States exported 531,000 barrels daily and imported 149,000; by 1947, the favorable balance had shifted: 436,000 barrels were exported daily, and 453,000 barrels imported. Since then, increased world tensions, the Korean War, and military necessity have intensified concern over domestic supply and the importation problem.

Admiral Arthur Radford, chairman of the joint chiefs of staff, recently listed three sets of conditions under which the military establishment is called upon to function:

1. A so-called cold war, in which world tensions of greater or less degree may continue for a prolonged period of time.

2. Limited, peripheral or "brush-fire" type of war, the onset of which might occur in a number of widely separated areas of the globe; and, lastly,

3. General, or world war.[16]

The latter certainly would give rise to petroleum shortages and is under continuous study by the National Security Council.

[15] *Ibid.,* 658.

[16] "Sources of Supply of Oil and Other Petroleum Products for West Coast of the United States for Military and Other Requirements," testimony of Admiral Radford on January 25, 1956, at Hearings before a Subcommittee for Special Investigation of the Committee on Armed Services, 84 Cong., *House Resolution 112* (1956), Part 2.

From the day of the attack on Pearl Harbor, December 7, 1941, to June 30, 1946, the Defense Supplies Corporation spent $2,-202,267,571 for 131,913,663,862 gallons of 100-octane aviation gasoline. During the fiscal year alone, ending June 30, 1956, a year of normal peacetime operation, the Department of Defense spent more than $1,000,000,000 for fuels and lubricating oils, and approximately $1,200,000,000 for this purpose during the year ending June 30, 1957. These figures perhaps become more significant when it is recalled that even the costs of fuel and lubricating oil used per flying hour have increased, although cheaper fuels are now used. During World War II, the cost of fuel for one flying hour of the B–25 bomber was $29; for the B–26, $31; now, it costs $377 for enough fuel and oil to keep a B–52 bomber aloft one hour.[17]

The National Security Council has had under consideration the advisability of constructing a thirty-inch crude oil pipe line from El Paso to Los Angeles—over 1,000 miles—to provide oil to the West Coast. It is 5,000 miles by water from the Gulf Coast–Southwest to the West Coast. It is 4,400 water miles from the Caribbean and Central America. It is 11,400 miles from Saudi Arabia. Council members know that the Russians have over 400 snorkel-type submarines, that are large and fast; they know that the Nazis in World War II with only 150 submarines in limited use off "torpedo alley"—the sea lanes of the East Coast —sank 54 of our 59 tankers. These are some of the reasons why the importation problem has become interrelated with problems of national security in addition to its effect upon the national economy during normal peacetime operations.

The cumulative production of oil since it was first produced commercially up to 1950 has been estimated to be 61,728,000,-000 barrels. Of this amount, the United States produced 63.2

[17] "Department of the Air Force, Appropriations for 1959," Hearings before the Subcommittee of the Committee on Appropriations, 85 Cong., 2 sess. (1958), 614–15.

per cent; the Western Hemisphere, 78.2 per cent. But estimated reserves in the United States amounted to only 26 per cent; reserves of the Eastern Hemisphere, principally the Middle East, amounted to 52.6 per cent.[18]

Important oil-producing areas outside of the United States and Russian-controlled territory are Venezuela, the Middle East, and Indonesia. Mexico also produces a substantial amount of oil, and Canadian production helps to supplement this country's supply. American companies produce approximately 66 2/3 per cent of Venezuela's output, more than 50 per cent of the Middle East output, and about 38 per cent of the output of Indonesia. In 1949, American companies produced about 50 per cent of the oil produced outside the United States and Eastern Europe and controlled reserves of about 7,500,000,000 barrels in South America, 16,000,000,000 barrels in the Middle East, and probably 1,000,000,000 barrels elsewhere—about 45 per cent of the world total outside the United States—which is slightly less than the total domestic reserves.[19] It has been estimated that investments in foreign concessions, production, refining, and distributing facilities of the industry accounted for 26 per cent of the total United States private investments abroad by the end of 1948, and for approximately 80 per cent of the new investments made between 1948 and 1952. More than $4,500,000,000 were invested abroad during that five-year span.

An example of American activity in a foreign concession may be illustrated by the Arabian American Oil Company ("Aramco"). It is owned by four American oil concerns: Standard Oil of New Jersey, Standard Oil of California, and the Texas Company each own 30 per cent; the Socony Mobil Oil Company owns 10 per cent. A so-called "50-50 agreement" was made in December, 1950, which determines the company's and Saudi

18 "Petroleum Study," *loc. cit.*, 216.
19 "Statement of Willard L. Thorpe, assistant secretary of state for economic affairs, April 5, 1950," in *ibid.*, 217.

Arabia's shares: Saudi Arabia receives from Aramco in a combination of "taxes, rentals, per ton royalties and other levies an amount equal to the petroleum producer's net income after all charges and taxes."[20] This type of agreement—viz., an equal division of profits between the operator and the government—is maintained in most foreign concessions. Landowners in some South American countries receive a royalty which varies from 1 to 4 per cent; in Canada, some of the producing properties are on land granted years ago to homesteaders or to the Canadian Pacific Railroad with all mineral rights. Although world royalty rates vary widely from country to country, it is an accepted practice outside of the United States for the government not only to share equally with the operator any profits from oil production but also to benefit from leasing agreements and royalty payments.[21]

The greatest foreign production since World War II has occurred in the Middle East. Production in Kuwait amounted to 32,217 barrels daily in 1946; in 1955, daily production was 1,091,763 barrels from 185 wells, more than one-third of the total production in the Middle East, Saudi Arabia, which accounts for slightly less than one-third of the total production, had an annual production of only 19,777 barrels in 1936. This increased to 4,530,492 barrels in 1942, to 59,943,766 barrels in 1946, and to 360,923,384 barrels in 1956, an average daily production of 986,129 barrels from 164 wells. It has recently been estimated that production from Middle East fields should amount to about 8,000,000 barrels daily by 1965.[22]

[20] The agreement appears in "Sources of Supply of Oil and Other Petroleum Products for West Coast of the United States for Military and Other Requirements," loc. cit.

[21] "World Royalty Rates," World Petroleum, Vol. XXVII, No. 11 (October, 1956), 65.

[22] Lomax, "Expansion in Kuwait," loc. cit., 55; also, "1956 Report of Operations to the Saudi Arab Government by the Arabian American Oil Company."

The following table shows the extent of foreign interests, including American oil companies, in the Middle East:[23]

PRODUCTION, BY OWNERSHIP, IN MIDDLE EAST
DURING FIRST SIX MONTHS (1956)
(LONG TONS IN MILLIONS)

	British	Royal Dutch Shell	French	United States	Gulbenkian
Iraq	3.97	3.97	3.97	3.97	.83
Qatar	.60	.60	.60	.60	.15
Iran	4.86	1.70	.73	4.86	——
Kuwait	14.40	——	——	14.40	——
Aramco	——	——	——	25.75	——
Neutral Zone	——	——	——	.75	——
Bahrein	——	——	——	.75	——
	23.83	6.27	5.30	51.08	.98
Per cent of ownership	27.2	7.3	6.0	58.4	1.1

Bearing in mind the great flood of oil produced from wells in the flush fields of the Middle East, we find little wonder that Ernest O. Thompson, railroad commissioner of Texas, when addressing a Congressional committee on oil imports, exclaimed: "How can a Texas twenty barrel allowable compete with a five thousand barrel well!"[24] Producing wells in the United States

---

[23] E. Lawson Lomax, "London Letter," *World Petroleum,* Vol. XXVII, No. 11 (October, 1956), 76. Although refining is increasing in the Middle East—the Ras Tanura refinery of Aramco processed an average of 203,210 barrels of crude oil a day in 1955, a total for the year of 71,171,745 barrels, or about one-fifth of the total production—practically all the crude oil processed in Western Europe comes from the Middle East. More than one-half of all Middle East crude—over 1,000,000 barrels a day—is normally carried through the Suez Canal.

[24] Hearings before the House Committee on Interstate and Foreign Commerce, February 11, 1953, 83 Cong., 1 sess. (1953), 690. In the "1955 Report of Operations to the Saudi Arab Government by the Arabian American Oil Company," the company estimated reserves at the end of the year to be 35,000,000,000 barrels, which is almost 5,000,000,000 barrels more than the proved reserves of the United States. According to an A.P.I. news release of March 15, 1956, the proved reserves in this country were estimated to be 30,012,000,000 barrels.

average thirteen barrels of oil a day, and in 1955 the average gross increase in proved oil reserves per well drilled was only 60,500 barrels, or only two-thirds of the total amount drilled during 1951–53.[25]

A recently published report on domestically drilled wells in unproved territory, that is, prospective new field exploratory efforts, revealed the following: Only 1 well in 11.9 discovered any oil or gas; 1 in 14.1 discovered oil; only 1 in 42 discovered a field with reserves of 1,000,000 barrels; 1 in 158 discovered a field with 10,000,000 barrels; 1 in 427 discovered one with 25,-000,000 barrels; and 1 in 706 discovered a field with reserves of 50,000,000 barrels.[26]

With proved oil reserves increasing in other parts of the world, and with the United States finally entering a period when consumption exceeds domestic supply, it is inevitable that the Committee of Economic Advisors to the President, the National Security Council, and the Congress must establish and maintain policies that will assure an adequate supply.

Imports of oil into the United States have exceeded exports since 1948, and most of the imports originate in foreign concessions owned by subsidiaries or affiliates of American oil companies.[27] The principal sources are Venezuela, Canada, the Dutch West Indies, Colombia, and the Middle East. The United States has outstanding "most-favored-nation" agreements with respect to imports, based upon the Venezuelan agreement of 1940. The tariff amounts to ten and one-half cents a barrel on

---

[25] Herbert Hoover, Jr., "Petroleum Imports," *The Oil and Gas Compact Bulletin,* Vol. XV, No. 1 (June, 1956), 16.

[26] F. H. Lahee, "How Many Fields Really Pay Off?" *The Oil and Gas Journal,* Vol. LIV, No. 72 (September 17, 1956), 369–71. Mr. Lahee's study was made of wells drilled in unproved territory during 1944–49, inclusive, and of the experience of those successful six years from 1949 to 1955; of those drilled in 1949, and of the six years from 1948 to 1954; of those drilled in 1948, etc. A total of 20,478 wells were drilled; 18,169 of them were dry, 1,701 produced oil, and 608 were gasers.

[27] "Petroleum Study," *loc. cit.,* 32.

oil imported, so long as the volume is less than 5 per cent of the total amount of refined products of the United States during the previous calendar year; if imports exceed this figure, the tariff is doubled.[28]

The Independent Petroleum Association filed an application with the United States Tariff Commission to re-examine its trade agreements and to consider an upward revision in tariff rates on oil, but the request was denied on May 3, 1949. About the same time, the National Petroleum Council, an advisory body to the secretary of the interior, formulated and recommended its plan for a national oil policy with respect to imports, in these words:

> The nation's economic welfare and security requires a policy on petroleum imports which will encourage exploration and development efforts in the domestic industry and which will make available a maximum supply of domestic oil to meet the needs of the nation.
>
> The availability of petroleum from domestic fields produced under sound conservation practices, together with other pertinent factors, provides the means for determining if imports are necessary and the extent to which imports are desirable to supplement our oil supplies on a basis which will be sound in terms of the national economy and in terms of conservation.
>
> The implementation of an import policy, therefore, should be flexible so that adjustments may readily be made from time to time.
>
> Imports in excess of our economic needs, after taking into account domestic production in conformance with good conservation practices and within the limits of maximum efficient rates of production, will retard domestic exploration and the development of new oil fields and the technological progress in all branches of the industry which is essential to the nation's economic welfare and security.

[28] *Ibid.*, 199, 202.

President Dwight D. Eisenhower established the Committee on Energy Supplies and Resources in July, 1954. It was composed of seven cabinet members under the chairmanship of Arthur S. Flemming, director of the Office of Defense Mobilization. The committee was directed by the President to study and evaluate all factors pertaining to the continued development of energy supplies and resources in the United States "with the aim of strengthening the national defense, providing orderly industrial growth, and assuring supplies for our expanding national economy and for any future emergency." Under this policy, domestic and military needs for oil and oil products have been given top consideration by the committee.

The need to draw upon foreign sources of supply is inevitable under these criteria. However, they do not conflict with the policy toward imports formulated by the National Petroleum Council in 1949, that oil from abroad should be available to the extent that it may be needed to supplement domestic supplies.

The President's Committee on Energy Supplies and Resources recommended in February, 1955, that petroleum imports should be held to their 1954 relationship to domestic crude oil production, and warned that if imports exceeded that ratio they would endanger national security. The 1954 ratio of imports to United States production was 16.6 per cent, or about 700,000 barrels daily. The committee hoped that voluntary restraints would be imposed by oil-importing companies to maintain the set ratio, and rejected the plea of proponents who advocated either higher tariff rates or the imposition of a quota system.

Herbert Hoover, Jr., undersecretary of state, explained in June, 1956, why discriminatory measures were not being sanctioned by the present administration.[29] He said that raising duties sufficiently to restrain imports would have resulted in

[29] Hoover, "Petroleum Imports," an address delivered at the 1956 mid-year meeting of the Interstate Oil Compact Commission held in Dallas, Texas, printed in *loc. cit.*, 15–17.

barring imports from certain relatively high-cost producing areas which are particularly important to our national defense. He pointed out that nondiscrimination in commercial treaties has historically been the policy of this country; it is the "most-favored-nation" concept of world trade. Quotas were rejected because they tend to freeze sources of supply, and "such a method would have placed shackles on an industry whose dynamic qualities should be fostered rather than hampered." Furthermore, according to Hoover, quotas would ultimately lead to governmental price-fixing and to further controls and regimentation.

The administration's point of view, as expressed by Hoover, has not remained unchallenged.[30] Some exception has been taken

[30] Warwick M. Downing, the official representative to the Interstate Oil Compact Commission from Colorado, took strong exception to the voluntary restraint program of the oil-importing companies, which was the system approved by the President's Committee. Judge Downing, in the course of the open forum held at Dallas after Hoover's speech, stated: "I don't think this meeting should close until I, for one, state to Mr. Hoover that I disagree emphatically with his argument. It looks to me as if his speech was an attack on free government. His argument is that government should not control big business. Big business should be free, under the theory of private enterprise, to do as it pleases, no matter how harmful it may be to the rest of us. With that concept I disagree, and wish to express my disagreement here.

"I want to express my appreciation of the job the Defense Mobilization Commission has done and what it is trying to do, but I think, after all, the basic question is the right of a free people to protect themselves in their own best interests. . . . Free enterprise has never meant that business should be free of Government control, where such control is necessary in the public interest. As proof of this, I cite the fact that every big company in America, and I think, practically every operator in America, believes in governmental controls by the states to insure conservation of oil and gas."

Downing's speech was quoted in *The Oil and Gas Compact Bulletin,* Vol. XV, No. 1 (June, 1956), 18.

Judge Downing of Denver, Colorado, has devoted a greater span of service to the oil and gas conservancy movement than has any other active participant. He represented the governor of Colorado at the Colorado Springs conference of governors and their representatives from the oil-producing states, which was called by Ray Lyman Wilbur, secretary of the interior, in 1929. Out of that meeting was formed the Oil States Governors' Committee; Judge Downing has held an active and continuous membership. He has served continuously as Colorado's representative to the Interstate Oil Compact Commission ever since its inception. At Dallas, Texas, in February, 1935, when Governor James Allred of Texas and Governor E. W. Marland of Oklahoma led forces in a disagreement on the to the committee's policy, which excludes oil from Canada and Venezuela from the voluntary compliance program. These countries were placed in a separate category because of national defense implications. In the event of a national emergency, oil from the Western Hemisphere countries will always be recognized as our safest supplemental source of petroleum. It will be recalled that in the first few months of World War II German submarines destroyed most of the tankers traveling the sea lanes from the Gulf Coast area and the Caribbean to the East Coast; deliveries in December, 1941, amounted to 1,421,000 barrels daily; in May, 1942, only 173,000 barrels; a year later, 63,000 barrels a day. That is why, from the standpoint of national security, the National Security Council and other federal agencies and the oil industry have had under advisement the construction of a thirty-inch crude oil pipe line from El Paso to the West Coast. Oil from the Middle East now supplements the California supply, as does Canadian oil. In 1955, Canada exported to our West Coast 23,070 barrels daily, but during the first quarter of 1956 this had been increased to 120,000 barrels daily.

It has been estimated that oil from the Persian Gulf can be delivered to California for $3.34 a barrel ($1.97 plus the tanker charge of $1.37), and oil from the eastern Mediterranean can be delivered for $3.76 ($2.39 plus the tanker charge of $1.37), whereas oil from West Texas, if the pipe-line development went through, could be delivered for $3.19 ($2.69 plus the pipe-line charge of $0.50).[31] At the present time, large natural gas transmission lines are moving 1,800,000,000 cubic feet of natural gas each day from West Texas and New Mexico to

terminology of the proposed compact, Judge Downing took the lead in working out a compromise. He has served as chairman of the Oil and Gas Conservation Commission of Colorado since 1932, when the agency was first established. He has been an active director or officer of the American Petroleum Institute, the Independent Petroleum Association of America, and the Rocky Mountain Oil and Gas Association.

[31] "Sources of Supply of Oil and Other Petroleum Products for West Coast of the United States for Military and Other Requirements," *loc. cit.*, Part 2, p. 857.

California, and 300,000,000 cubic feet each day to the Pacific Northwest. The total of 2,100,000,000 cubic feet represents an oil equivalent of 350,000 barrels of crude oil a day.

The United States has encouraged the building of the Trans-Mountain Pipeline from Alberta, Canada, to the Pacific Coast, in order to provide that rapidly growing and strategically important area with an assured source of crude oil. Natural gas, likewise, is being piped from Canada into the Pacific Northwest, while below the Texas border, Petroleos Mexicanos will soon begin the daily delivery of 115,000,000 cubic feet of natural gas to the Texas Eastern Transmission Corporation. This company is building a 422-mile, thirty-inch pipe line from the Río Grande to Beaumont, Texas; the Mexican company has a plant at Reynosa, in eastern Mexico, which delivers 300,000,000 cubic feet daily, and has reserves estimated to amount to 840,000,000,000 cubic feet. The United States looks to its northern and southern neighbors to supplement its fuel needs in the event of a national emergency.

Despite the dependence of the present administration on a voluntary restraint policy by oil-importing companies, nineteen associations of independent oil and gas producers in August, 1956, presented a twenty-six-point statement to the President's Committee, in which they pointed out that imports were exceeding 20 per cent of the United States' production, instead of the 1954 ratio of 16.6 per cent. They warned that excessive importation would result in reduced drilling activity in this country, and would thus lessen the development of undiscovered oil reserves. Their principal protest was that "it is not sound national policy, in matters involving national security, to rely on voluntary actions of importing companies."

The independents presented evidence that imports in 1955 exceeded the 1954 yardstick by 120,000 barrels daily, and by more than 200,000 barrels daily during the first half of 1956. By late October, imports were exceeding the standard by 400,000 barrels daily.

Director Flemming made three pleas to the importers during the year to restrict the volume of crude oil shipped to the United States; in August, he warned fifty of the importing companies that they must voluntarily curtail the flow of oil or face possible compulsory curtailment. The seizure by Egypt of the Suez Canal in July pointed up the danger of interdependence on foreign oil sources, and the blocking of the canal in November after the French-British military action temporarily lessened the need for concern over oil imports into the United States. The blocking of the Suez Canal stopped the flow of more than 1,000,000 barrels of crude oil daily from the Persian Gulf area to western Europe and diverted, for the moment, the resentment of domestic producers toward oil importers.[32] Because of the Suez crisis, Flemming set up a Middle East Emergency Committee in November, 1956, to work out tanker schedules to provide 850,000 barrels of Gulf Coast and Mid-Continent oil daily to Western Europe. Great Britain and the other countries of Western Europe imposed severe restrictive policies on domestic and industrial use of oil and by-products.

The increased demand for oil from the Southwestern states was easily met by increasing the daily allowable production. The Suez crisis ended in March, 1957; a bridge of tankers carrying American oil had, again, been a major factor in saving Western Europe from apparent disaster.

Effects of the aftermath on domestic production, however, are still being felt. Texas, for example, in March, 1957, was permitting eighteen producing days, or an allowable of 3,821,-426 barrels of oil a day. Thirteen months later, producing days were down to eight, and daily allowable production was 2,444,571

---

[32] Trade publications for July to November, 1956, regularly carried articles on imports of oil into the United States. Typical are: "Texan Tells Congress High Imports Jeopardizing Security," *Texas Oil Journal,* Vol. XXIII, No. 6 (October, 1956), 16–24; "Importers Get Stern New Warning of Defense Mobilizer," *Texas Oil Journal,* Vol. XXIII, No. 5 (September, 1956), 13–14; and, "Suez Canal Seizure Points Danger in U. S. Oil Imports," *ibid.,* 12.

barrels, a cutback of 1,376,855 barrels daily. This represents the lowest calendar-day allowable since June, 1950, when there were 119,000 producing wells in Texas; now there are 182,500, an increase of 63,500.[33]

Curtailments in exploration and production directly affect income and employment in the oil-producing states among geophysical firms, drilling contractors, and equipment, supply, and service companies. The impact gravitates back to manufacturing concerns; it is most apparent along the main streets of towns dependent upon the oil economy, and reduced production directly affects income from oil royalty and state revenue.

It has been estimated, for example, that royalty payments in 1957 for West Texas alone amounted to $258,000,000, but if allowable production in the region is not increased over the present amount, royalty owners will receive $103,000,000 less this year. Taxes on oil and gas operations amount to approximately one-third of the total property tax paid to Texas' 254 counties and, according to the Texas Mid-Continent Oil and Gas Association, the industry pays 69 per cent of all business and property taxes collected by the state. A decline in revenue from this source affects all agencies of the state government. Oil-producing states are dependent upon revenue from the industry, through the administration of severance tax or gross production tax laws. With production in Texas for April, 1958, set 32 per cent below the rate of March, 1957, and the allowable production for other states reduced 20 per cent, domestic producers were hopeful that importing companies would not obtain a greater proportion of the market.

In mid-summer of 1958, the Cabinet Fuels Committee set the import ratio at 12 per cent of the production east of the Rocky Mountains and estimated that this would not discourage exploration and production to an extent affecting national se-

[33] Robert J. Enright, "What 8 Days Will Mean to Texas," *The Oil and Gas Journal,* Vol. LVI, No. 13 (March 31, 1958), 47–49.

curity. Allocations were made and set for a twelve-month period for importers to Gulf and Atlantic coastal points; no limitation on imports was found necessary for the West Coast.

In January, 1958, the voluntary program was extended to the West Coast. With the completion in April of a sixteen-inch, 750-mile-long pipe line from the Four Corners region—where the states of New Mexico, Arizona, Utah, and Colorado meet— to the Los Angeles area, the West Coast is no longer dependent upon crude from the Middle East. This points up another elementary fact in the controversy between domestic producers and foreign companies competing for the American market. Tankers may transport cheap foreign oil, place it on barges, and compete successfully for the Detroit, St. Louis, Chicago, and Cincinnati markets. But they cannot transport the natural gas which is necessary and useful for domestic and industrial users. Approximately 10,000,000,000,000 cubic feet of natural gas were delivered to 27,000,000 customers in 1956, and an estimated 6 per cent increase was made the following year.

Natural gas in 1957 accounted for nearly 34 per cent of the domestic petroleum market. Since new reserves are found through oil exploration and since natural gas cannot be transported across the oceans by tanker, any lessening of exploratory activity diminishes known reserves.

President Eisenhower in March, 1958, issued an executive order pertaining to the importation of foreign oil from April 1 to September 1.[34] The order permitted the Cabinet Fuel Committee to examine and evaluate its effect at any time. The 12 per cent ratio for imports east of the Rocky Mountains was retained, with an over-all limitation for daily importation set at 713,000 barrels.

Certificates of compliance have been issued to importing firms, and individual quotas have been set. An importing firm is not in compliance if it exceeds its established allocation in any

[34] *The Oil and Gas Journal,* Vol. LVI, No. 13 (March 31, 1958), 50–51.

three consecutive months; noncompliance may bring cancellation of the certificate, and, perhaps even more compelling, forfeiture of the right to supply any of the government's petroleum needs—a billion-dollar business. Government contracts for petroleum products now include this clause: "Contractor agrees that during the contract period he will comply in all respects with the voluntary oil program."

For more than thirty years, independent producers have appeared before committee hearings of the successive sessions of Congress to urge legislation for a quota system which would be strictly administered with a penalty for noncompliance, or for higher duties on imported oil and oil products. They have had strong but not numerous enough advocates in the congresses.[35] Perhaps, with government purchases a factor in marketing, the present voluntary program for importers can be workable and will thus make mandatory legislation unnecessary. Throughout the era of conservation, however, voluntary controls broke down because they could be violated with impunity, and it became necessary to accomplish objectives through legislation and court decree. Proponents in Congress for a strict quota system or higher duties on petroleum and by-products now watch with interest the voluntary-compliance plan.

[35] Congressman Frank Ikard of Wichita Falls, Texas, with enthusiastic support from the Independent Petroleum Association of America, sought to amend the Reciprocal Trade Act of 1958, to limit the importation of crude and crude products to their 1954 levels. The House Ways and Means Committee rejected the amendment by a vote of 18 to 7. Ikard proposed that the allocation for crude oil imports be reduced from 940,000 to 670,000 barrels daily, and for crude products, from 575,000 to 410,000 barrels daily.

APPENDIX

TEXT OF THE COMPACT
AN INTERSTATE COMPACT TO
CONSERVE OIL AND GAS

ARTICLE I

This agreement may become effective within any compacting state at any time as prescribed by that state, and shall become effective within those states ratifying it whenever any three of the States of Texas, Oklahoma, California, Kansas and New Mexico have ratified and Congress has given its consent. Any oil producing state may become a party hereto as hereinafter provided.

ARTICLE II

The purpose of this Compact is to conserve oil and gas by the prevention of physical waste thereof from any cause.

ARTICLE III

Each state bound hereby agrees that within a reasonable time it will enact laws, or if laws have been enacted, then it agrees to continue the same in force, to accomplish within reasonable limits the prevention of:

(a) The operation of any oil well with an inefficient gas-oil ratio.
(b) The drowning with water of any stratum capable of producing oil or gas, or both oil and gas in paying quantities.

(c) The avoidable escape into the open air or the wasteful burning of gas from a natural gas well.

(d) The creation of unnecessary fire hazards.

(e) The drilling, equipping, locating, spacing or operating of a well or wells so as to bring about physical waste of oil or gas in the ultimate recovery thereof.

(f) The inefficient, excessive or improper use of the reservoir energy in producing any well.

The enumeration of the foregoing subjects shall not limit the scope of the authority of any state.

ARTICLE IV

Each state bound hereby agrees that it will, within a reasonable time, enact statutes, or if such statutes have been enacted, then it will continue the same in force, providing in effect that oil produced in violation of its valid and/or gas conservation statutes or any valid rule, order or regulation promulgated thereunder, shall be denied access to commerce; and providing for stringent penalties for the waste of either oil or gas.

ARTICLE V

It is not the purpose of this Compact to authorize the states joining herein to limit the production of oil or gas for the purpose of stabilizing or fixing the price thereof, or create or perpetuate monopoly, or to promote regimentation, but is limited to the purpose of conserving oil and gas and preventing the avoidable waste thereof within reasonable limitations.

ARTICLE VI

Each state joining herein shall appoint one representative to a commission hereby constituted and designated as the Interstate Oil Compact Commission, the duty of which said Commission shall be to make inquiry and ascertain from time to time such methods, practices, circumstances and conditions as may be disclosed for bringing about conservation and the prevention of physical waste of oil and gas, and at such intervals as said Commission deems beneficial it shall report

256

its findings and recommendations to the several states for adoption or rejection.

The Commission shall have power to recommend the coordination of the exercise of the police powers of the several states within their several jurisdictions to promote the maximum ultimate recovery from the petroleum reserves of said states, and to recommend measures for the maximum ultimate recovery of oil and gas. Said Commission shall organize and adopt suitable rules and regulations for the conduct of its business.

No action shall be taken by the Commission except: (1) By the affirmative votes of the majority of the whole number of the compacting states, represented at any meeting, and (2) by a concurring vote of a majority in interest of the compacting states at said meeting, such interest to be determined as follows: Such vote of each state shall be in the decimal proportion fixed by the ratio of its daily average production during the preceding calendar half-year to the daily average production of the compacting states during said period.

ARTICLE VII

No state by joining herein shall become financially obligated to any other state, nor shall the breach of the terms hereof by any state subject such state to financial responsibility to the other states joining herein.

ARTICLE VIII

This Compact shall expire September 1, 1937. But any state joining herein may, upon sixty (60) days notice, withdraw herefrom.

The representatives of the signatory states have signed this agreement in a single original which shall be deposited in the archives of the Department of State of the United States, and a duly certified copy shall be forwarded to the Governor of each of the signatory states.

This Compact shall become effective when ratified and approved as provided in Article I. Any oil producing state may become a party hereto by affixing its signature to a counterpart to be similarly deposited, certified and ratified.

Done in the City of Dallas, Texas, this sixteenth day of February, 1935.

BIBLIOGRAPHY

1. *Government Publications*

(a) *Congressional Documents*

Hearings before the Subcommittee of the Committee on Appropriations, House of Representatives, "Department of the Air Force, Appropriations for 1959," 85 Cong., 2 sess. (1958).

Hearings before a Special Senate Committee Investigating Petroleum Resources, 79 Cong., *Senate Resolution 36* (1946). The subjects included are: "The Independent Petroleum Company," "Investigation of Petroleum Resources," "A Preliminary Report of Its National Oil Committee to the Petroleum Industry War Council," "Report on Petroleum Economics by the Petroleum Industry War Council," "Wartime Petroleum Policy under the Petroleum Administrator for War," and "Wartime Petroleum Supply and Transportation."

Hearings before the Special Senate Committee to Study Problems of American Small Business, 80 Cong., *Senate Resolution 20* (1948).

Hearings before a Subcommittee of the House Committee on Interstate and Foreign Commerce, 73 Cong., *House Resolution 441* (1934); 75 Cong., *Senate Resolution 790* and *House Resolution 5366* (1937); 76 Cong., *House Resolutions 290 and 7372* (1939–40); 77 Cong., *House Resolutions 290, 15, and 118* (1941–42); and "Petroleum Study," 81 Cong., *House Resolutions 107 and 6047*, and *House Joint Resolution 423* (1956).

Hearings before the House Committee on Interstate and Foreign Commerce, 81 Cong., *House Resolutions 107 and 6047*, and *House*

Joint Resolution 423 (1950); "Interstate Oil and Gas Compact," 82 Cong., *House Joint Resolutions 206 and 211* (1951); and 83 Cong., 1 sess. (1953).

Hearings before a Subcommittee of the Senate Committee on Banking and Currency, 81 Cong., 1 sess. (1949).

Hearings before a Senate Subcommittee of the Committee on Finance, 75 Cong., *Senate Resolution 790* (1937).

Hearings before the Senate Committee on Interior and Insular Affairs, the Subcommittee on Minerals, Materials, and Fuels, 83 Cong., *Senate Resolution 143* (1954).

Hearings before a Subcommittee for Special Investigations of the Committee on Armed Services, on "Sources of Supply of Oil and Other Petroleum Products for West Coast of the United States for Military and Other Requirements," 84 Cong., *House Resolution 112* (1956).

Hearings before the Temporary National Economic Committee, on "Investigations of the Concentration of Economic Power," 76 Cong., 2 sess., *Public Resolution 113* (1940). (Parts 14–16 pertain to the petroleum industry.)

"Internal Revenue Act of March 3, 1865," United States Statutes at Large.

Miller, H. C. and G. B. Shea. "Recent Progress in Petroleum Development and Production," Hearings on *House Resolutions 290 and 7372*, 76 Cong., (1939–40).

"Oil Concessions in Foreign Countries," a report of the Secretary of State relative to securing oil concessions for American citizens between this and foreign governments, 68 Cong., 1 sess., *Sen. Doc. 97* (1924).

"Oil Supply and Distribution Problems," 81 Cong., 1 sess., *Senate Report 25, The Wherry Report* (1949).

Peckham, S. F. "Report on the Production, Technology, and Uses of Petroleum and Its Products," 47 Cong., 2 sess., *House Misc. Doc. 42*, Part 10, Department of the Census (1884).

Stowell, S. H. "Petroleum," in *Mineral Resources of the United States.* United States Geological Survey, Department of Interior, 48 Cong., 1 sess., *House Misc. Doc. 75* (1883).

(b) *Articles and Reports*

Blatchley, Raymond S. "Waste of Oil in the Mid-Continent Fields." *Technical Paper 45,* Department of Interior, Bureau of Mines. Washington, 1913.

Eliot, Charles B. "Petroleum Industry in the Southwest." Commercial Survey of the Gulf Southwest. Department of Commerce, Series No. 44, Part 2. Washington, 1913.

Federal Oil Conservation Board. Complete Records of Public Hearings, February 10–11, 1926. Washington, 1926.

Federal Oil Conservation Board, Reports I–V to the President of the United States. Washington, 1929–32.

Fowler, H. C. "Development of the American Petroleum Industry, 1914–18: Exploration, Drilling, Production, and Transportation." Information Circular, Department of Interior, Bureau of Mines. Washington, 1941.

Frey, John W. and H. Chandler Ide (editors). *A History of the Petroleum Administrator for War, 1941–45.* Washington, 1946.

Heggem, A. G. and J. A. Pollard. "Drilling Oil Wells by the Mud-Laden Fluid Method." *Technical Paper 68,* Department of Interior, Bureau of Mines. Washington, 1914.

Kirwan, M. J. "Effects of Extraneous Gas on the Productivity of Oil Wells in the Lyons-Quinn Field of Oklahoma," Department of Interior, Bureau of Mines. Washington, 1924.

Kraemer, A. J. "Development in Petroleum Refining Technology in the United States, 1914–19." Information Circular, Department of Interior, Bureau of Mines. Washington, 1941.

McMurray, William F. and James O. Lewis. "Underground Wastes in Oil and Gas Fields and Methods of Prevention." *Technical Paper 130,* Department of Interior, Bureau of Mines. Washington, 1914.

Mineral Yearbook: Fuels. Department of Interior. Washington, 1956. "Orders Relating to Oil and Gas Conservation Practices in Oklahoma," in *Reports VIII–L of the Corporation Commission of the State of Oklahoma.* Oklahoma City, 1916–57.

Petroleum Administration Board. *Operations of the New Pool Plans of Orderly Development under the Code of Fair Competition for the Petroleum Industry.* Washington, 1936.

Pollard, J. A. and A. G. Heggem. "Mud-Laden Fluid Applied to Well-Drilling." *Technical Paper 66,* Department of Interior, Bureau of Mines. Washington, 1914.

Reports I and II of the Attorney General, pursuant to Section 2 of the Joint Resolution of July 28, 1955, consenting to an Interstate Compact to Conserve Oil and Gas. Washington, 1956–57.

2. *Newspapers*

Advertiser (Nowata, Oklahoma).

Derrick (Oil City, Pennsylvania).

Daily Oklahoman (Oklahoma City, Oklahoma).

Herald (Titusville, Pennsylvania).

Morning Herald (Titusville, Pennsylvania).

Register (Oil City, Pennsylvania).

Venango Spectator (Franklin, Pennsylvania).

3. *Books and Monographs*

American Bar Association. *Legal History of Conservation of Oil and Gas: A Symposium.* Chicago, 1939.

American Petroleum Institute. *Petroleum Facts and Figures.* Ninth Edition, New York, 1951.

————. *A Survey of Present Position of the Petroleum Industry, and Its Outlook toward the Future.*

Ball, Max W. *This Fascinating Oil Business.* New York, 1940.

Bartley, Ernest R. *The Tidelands Oil Controversy: A Legal and Historical Analysis.* Austin, Texas, 1953.

Bone, J. H. A. *Petroleum and Petroleum Wells.* Philadelphia, 1865.

Breeding, Clark W. and A. Gordon Burton. *Taxation of Oil and Gas Income.* Englewood Cliffs, New Jersey, 1954.

Buckley, Stuart E. (editor). *Petroleum Conservation.* Dallas, 1951.

Clark, James A. *Three Stars for the Colonel.* New York, 1954.

Connelly, William L. *The Oil Business as I Saw It: Half a Century with Sinclair.* Norman, 1954.

"Crocus." See Leonard, Charles C.

DeGolyer, E. (editor). *Elements of the Petroleum Industry.* New York, 1940.

The Derrick's Hand-Book of Petroleum. 4 vols. Oil City, Pennsylvania, 1898–1920.

Ely, Northcutt (compiler). *The Oil and Gas Conservation Statutes.* Washington, 1933.

Fanning, Leonard M. *The Rise of American Oil.* New York, 1948.

———— (editor). *The Shift of World Petroleum Power Away from the United States.* A report issued by the Gulf Oil Corporation in Pittsburgh, April 10, 1958.

Flaherty, John J. *Flowing Gold: The Romance of the Oil Industry.* New York, 1945.

Forbes, Gerald. *Flush Production: The Epic of Oil in the Gulf Southwest.* Norman, 1942.

Giddens, Paul H. *The Beginnings of the Petroleum Industry, Sources and Bibliography.* Harrisburg, Pennsylvania, 1941.

————. *The Birth of the Oil Industry.* New York, 1938.

———— (editor and compiler). *Pennsylvania Petroleum, 1750–1872. A Documentary History.* Titusville, Pennsylvania, 1947.

————. *Standard Oil Company (Indiana), Oil Pioneer of the Middle West.* New York, 1955.

Glasscock, Carl Burgess. *Then Came Oil: The Story of the Last Frontier.* Indianapolis, 1938.

Hardwicke, Robert E. *Antitrust Laws et al. v. Unit Operation of Oil and Gas Pools.* New York, 1948.

Henry, J. D. *Thirty-five Years of Oil Transport: The Evolution of the Tank Steamer.* London, 1907.

Henry, J. T. *The Early and Later History of Petroleum, with Authentic Facts in Regard to Its Development in Western Pennsylvania.* Philadelphia, 1873.

Hines, LeRoy H. *Unitization of Federal Lands.* Denver, 1953.

Hoffman, Leo J. *Voluntary Pooling and Unitization: Oil and Gas.* Dallas, 1954.

House, Boyce. *Oil Boom: The Story of Spindletop, Burkburnett, Mexia, Smackover, Desdemona, and Ranger.* Caldwell, Idaho, 1941.

Ickes, Harold L. *Fightin' Oil*. New York, 1943.

Interstate Oil Compact Commission. *Oil and Gas Production: An Introductory Guide to Production Techniques and Conservation Methods*. Norman, 1951.

James, Marquis. *The Texaco Story. The First Fifty Years, 1902–52*. New York, 1953.

Johnson, Arthur M. *The Development of American Petroleum Pipelines: A Study in Private Enterprise and Public Policy, 1862–1906*. Ithaca, New York, 1956.

Josephson, Matthew. *The Robber Barons: The Great American Capitalists, 1861–1901*. New York, 1934.

Kansas, A Guide to the Sunflower State. Compiled and written by the federal writers' project of the Work Projects Administration for the state of Kansas . . . Sponsored by the state Department of Education. New York, 1939.

Leeston, Alfred M. *Magic Oil: Servant of the World*. Dallas, 1951.

Leonard, Charles C. ("Crocus"). *The History of Pithole*. Pithole City, Pennsylvania, 1867.

Levorson, A. I. *Geology of Petroleum*. San Francisco, 1954.

McGee, Dean A. "Economics of Offshore Drilling in the Gulf of Mexico." A paper for presentation during the thirty-fourth annual meeting of the American Petroleum Institute in Chicago, November 9, 1954. Mimeographed copy, New York, American Petroleum Institute.

McLaurin, John J. *Sketches in Crude Oil*. Harrisburg, 1896.

McLean, John G. and Robert W. Haigh. *The Growth of Integrated Oil Companies*. Boston, 1954.

Mid-Continent Oil and Gas Association. *Handbook on Unitization of Oil Pools*. St. Louis, 1930.

Miller, H. C. *Functions of Natural Gas in the Production of Oil*. New York, American Petroleum Institute, 1929.

Morris, Edmund. *Derrick and Drill, Or an Insight into the Discovery, Development, and Present Condition and Future Prospects of Petroleum in New York, Pennsylvania, Ohio, West Virginia, etc.* New York, 1865.

Murphy, Blakely M. (editor). *Conservation of Oil and Gas: A Legal History, 1948.* Chicago, 1949.

Myers, Raymond M. *The Law of Pooling and Unitization, Voluntary-Compulsory.* New York, 1957.

Netschert, Bruce C. *The Future Supply of Oil and Gas: A Study of the Availability of Crude Oil, Natural Gas, and Natural Gas Liquids in the United States in the Period through 1975.* Baltimore, 1958.

Newton, J. H. (editor). *History of Venango County, Pennsylvania.* Columbus, Ohio, 1879.

People's Petroleum Company. *Prospectus.* New York, 1865.

Pettingill, Samuel H. *Hot Oil! The Problem of Petroleum.* New York, 1936.

Rister, Carl Coke. *Oil! Titan of the Southwest.* Norman, 1949.

Shuman, Ronald B. *The Petroleum Industry: An Economic Survey.* Norman, 1940.

Snider, L. C. *Oil and Gas in the Mid-Continent Fields.* Oklahoma City, 1920.

Stocking, George Ward. *The Oil Industry and the Competitive System: A Study in Waste.* New York, 1925.

Tarbell, Ida M. *The History of the Standard Oil Company.* 2 vols. New York, 1933.

Tate, A. Norman. *Petroleum and Its Products.* New York, 1863.

Watkins, Myron V. *Oil: Stabilization or Conservation.* New York, 1937.

Westcott, James H. *Oil: Its Conservation and Waste.* New York, 1928.

Wilson, Charles Morrow. *Oil across the World.* New York, 1946.

Youle, W. E. *Sixty-three Years in the Oil Fields.* Taft, California, 1926.

Zimmerman, Erich W. *Conservation in the Production of Petroleum: A Study in Industrial Control.* New Haven, Connecticut, 1957.

4. *Periodicals and Special Reports*

Second Annual Report to the Stockholders of the Central Petroleum Company, December 15, 1865. New York, 1865. A copy of this report is in the Drake Museum in Titusville, Pennsylvania.

National Petroleum News. Published weekly. (May 22, 1918; June 4, 1919.)

The Oil and Gas Compact Bulletin. Published quarterly by the Interstate Oil and Gas Compact Commission. (1941–58.)

The Oil and Gas Journal. Published weekly. (1935–58.)

The Oil Weekly. Published weekly. (1937–47.)

1955–56 Reports of Operations to the Saudi Arab Government by the Arabian American Oil Company. New York, 1956–57.

The Texas Oil Journal. Published monthly. (1956–58.)

World Oil. Published monthly. (1947–58.)

World Petroleum. Published monthly. (1943–58.)

5. *Articles*

Anon. "Biggest Year for Big Business," *Fortune,* Vol. LIV, No. 1 (July, 1956).

———. "The Big Gulf Gamble," *Fortune,* Vol. XXXVIII, No. 1 (July, 1948).

———. "The Fortune Directory of the 500 Largest U. S. Industrial Corporations," *Fortune* (a special insert), Vol. LIV, No. 1 (July, 1956).

———. "How President's Order Cuts Imports," *The Oil and Gas Journal,* Vol. LVI, No. 13 (March 31, 1958).

———. "Many Industries Have a Stake in Oil's Proration Battle," *Business Week* (January 28, 1931).

———. "The New Argonauts," *Time,* Vol. LXVIII, No. 6 (August 6, 1956).

———. "New Import Plan Drafted," *The Oil and Gas Journal,* Vol. LIV, No. 76 (October 15, 1956).

———. "World Royalty Rates," *World Petroleum,* Vol. XXVII, No. 11 (October, 1956).

Calvert, James W. "New Floating Tenders for Offshore Work," *World Petroleum,* Vol. XXVII, No. 11 (October, 1956).

DeGolyer, E. L. "How Men Find Oil," *Fortune* (August, 1949).

Enright, Robert J. "What 8 Days Will Mean to Texas," *The Oil and Gas Journal,* Vol. LVI, No. 13 (March 31, 1958).

Hoover, Herbert, Jr. "Petroleum Imports," *The Oil and Gas Compact Bulletin*, Vol. XV, No. 1 (June, 1956).

Lahee, F. M. "How Many Fields Really Pay off?" *The Oil and Gas Journal*, Vol. LIV, No. 72 (September 17, 1956).

Lomax, E. Lawson. "Expansion in Kuwait," *World Petroleum*, Vol. XXVII, No. 11 (October, 1956).

————. "London Letter," *World Petroleum*, Vol. XXVII, No. 11 (October, 1956).

Stalcup, H. M. "What the Oil Industry Is Doing about Conservation," *The Oil and Gas Journal*, Vol. XXXVII, No. 1 (May 19, 1938).

Thompson, Lieutenant General Ernest O. "The Railroad Commission of Texas, Oil and Gas Conservation Authority—The Legal Bases of Its Operation," *The Oil and Gas Compact Bulletin*, Vol. XV, No. 1 (June, 1956).

Torrey, Paul D. "Evaluation of United States Oil Resources as of January 1, 1956," *The Oil and Gas Compact Bulletin*, Vol. XV, No. 1 (June, 1956).

Wright, Muriel H. "First Oklahoma Oil Was Produced in 1859," *Chronicles of Oklahoma*, Vol. IV, No. 4 (December, 1926).

6. Cases

Bandini Petroleum Company v. *Superior Court*, 110 Cal. App. 123 (November, 1930).

Bandini Petroleum Company et al v. *Superior Court, Los Angeles, California et al.*, 284 U. S. 8 (November, 1931).

Barnard v. *Monongahela Natural Gas Company*, 216 Penn. St. 362 (October, 1906).

Brown v. *Vandergrift*, 80 Penn. St. 142 (November, 1875).

C. C. Julian Oil and Royalties Company v. *Capshaw et al.*, 145 Oklahoma 237 (October, 1930).

Champlin Refining Company v. *Corporation Commission of Oklahoma et al.*, 286 U. S. 210 (May, 1932).

Corzelius v. *Railroad Commission et al.*, 182 S. W. (2d) 412 (July, 1944).

Corzelius et al. v. *Harrell*, 186 S. W. 2d 961 (April, 1945).

Davenport v. *East Texas Refinery Company,* 127 S. W. (2d) 316 (April, 1939).

E. Constantin et al. v. *Lon Smith et al.,* 57 Fed. (2d) 227 (February, 1932).

H. F. Wilcox Oil and Gas Company v. *State et al.,* 162 Oklahoma 89 (February, 1933).

H. F. Wilcox Oil and Gas Company v. *Walker et al.,* 168 Oklahoma 355 (May, 1934).

Kansas Natural Gas Company v. *Haskell et al.,* 172 Fed. Report 545 (July, 1909).

Kelly v. *Ohio Oil Company,* 49 N. E. Report 399 (December, 1897).

MacMillan et al. v. *Railroad Commission of Texas et al.,* 51 Fed. (2d) 400–405 (July, 1931).

Murphy Oil Company v. *Burnet, Commissioner of Internal Revenue,* 287 U. S. 299 (December, 1932).

Ohio Oil Company v. *Indiana,* 177 U. S. 190 (April, 1900).

Panama Refining Company v. *Ryan et al.,* 293 U. S. 388 (January, 1935).

Patterson v. *Stanolind Oil and Gas Company,* 182 Oklahoma 155 (March, 1938); 305 U. S. 376 (January, 1939).

People v. *Associated Oil Company et al.,* 211 California 93 (December, 1930); 212 California 76 (March, 1931).

The Pipe-Line Cases, 234 U. S. 559 (June, 1914).

Scofield v. *Railway Company,* 43 Ohio State Reports 571 (January, 1885).

The Standard Oil Company of New Jersey et al. v. *The United States,* 221 U. S. 7 (May, 1911).

United States v. *Eason Oil Company,* 8 Fed. Supp. 365 (September, 1934); 79 Fed. (2d) 1013 (June, 1935).

United States v. *State of California,* 332 U. S. 19 (June, 1947).

United States v. *State of Louisiana,* 339 U. S. 699 (June, 1950).

United States v. *State of Texas,* 339 U. S. 707 (June, 1950).

Walls v. *Midland Carbon Company,* 254 U. S. 300 (December, 1920).

Waters-Pierce Oil Company v. *Texas,* 177 U. S. 28 (March, 1900).

West v. *Kansas Natural Gas Company,* 221 U. S. 229 (May, 1911).

Westmorland and Cambria Natural Gas Company v. *Dewitt et al.,* 130 Penn. St. 362 (November, 1889).

INDEX

THE OIL CENTURY

was set on the Linotype machine in 11-point Old Style No. 7
with 2 points of space between the lines. The title page and
initial letters in the chapter openings are set in Bodoni, a type
that goes well with the Old Style face. *The Oil Century* is printed
on an antique wove paper.

UNIVERSITY OF OKLAHOMA PRESS : NORMAN